Answer Engine Optimization

The 2026 Answer Engine Journal Guide to AI Visibility, SEO, GEO, and AEO

Zachary Handa · Georg Sinn

Demeter Academy Press · 2025

Answer Engine Optimization: The 2026 Answer Engine Journal Guide to AI Visibility, SEO, GEO, and AEO

Published by Demeter Academy Press, Los Angeles

For permissions, inquiries, and additional resources, visit:
www.DemeterAcademy.com

Interior Design & Typesetting: Demeter Academy Press
Printed in the United States of America

ISBN: 978-1-971093-01-7
First edition, 2025

Contents

How to Use This Guide ······ 7

Foundations of Answer Engine Optimization

Introduction to Answer Engine Optimization ······ 19

How Search Engines and LLMs Work ······ 24

Information Authority in the Age of AI ······ 28

Framing AEO Within Broader Search Intent ······ 31

Summary and Action Steps ······ 34

Strategic AEO Content Development

Creating and Scaling Pillar Content for AEO ······ 38

Structured Data and Technical Optimization ······ 44

Supporting Content Strategy and Internal Ecosystems ······ 48

Creating Content Designed for Specific Answer Contexts ······ 51

Asset Distribution and Knowledge Authority Building ······ 56

Summary and Action Steps ······ 60

Maximizing Visibility in AI and Search

Featured Snippets and Zero-Click Optimization ······ 65

Dominating People Also Ask (PAA) 68

Voice Search Optimization—Becoming the Spoken Answer 71

Getting Retrieved by LLMs and Generative AI 76

Influencing LLM Training Data Inclusion 79

Local Answer Engine Optimization—Winning the "Near Me" Query 83

Optimizing for Search in AI Interfaces 88

Channel-Specific Strategies 91

Part III Summary and AEO Visibility Checklist 95

Optimization, Measurement, and the Future

Monitoring AEO Performance Across Platforms 100

Debugging Ranking Drops and Indexing Issues 105

Iteration and Lifecycle Management of Assets 110

Ethical Considerations and Transparency in the AI Age 113

The Future of AEO—Strategic Foresight 116

The LLM SEO Toolkit—Tools to Operationalize Your Strategy 119

Final Checklist and AEO Maturity Model 124

Glossary 128

INTRODUCTION

How to Use This Guide

This book is built to be used, not just read.

Answer Engine Optimization (AEO) sits at the intersection of SEO, content strategy, analytics, and AI. Different readers will come to it with different strengths. This section will help you decide where to start, how to move through the material, and how to turn concepts into shipped work.

WHO THIS BOOK IS FOR

You'll get the most value from this guide if you are one of the following:

- **In-house SEO or content lead** responsible for organic visibility, content performance, or experimentation.
- **Agency strategist or consultant** helping clients adapt to AI search, SGE, and answer engines.
- **Product, growth, or marketing leader** who needs an executive-level view of how AI-driven discovery is changing strategy.
- **Founder, subject-matter expert, or publisher** building a durable authority presence across web, search, books, and AI interfaces.

You do not need to be a developer, data scientist, or LLM engineer. Where concepts lean technical, the focus stays on what to do in practice and which signals actually matter for being selected as an answer.

How the Book Is Structured

The guide is organized into four parts, each with a specific job:

Part I – Foundations of Answer Engine Optimization

What AEO is, how search engines and LLMs work, why authority signals matter, and how AEO differs from traditional SEO. This is where you get the mental model for "being the answer," not just "ranking for a keyword."

Part II – Strategic AEO Content Development

How to build pillar content, clusters, and supporting assets that are structured for retrieval and citation; how to use schema and internal ecosystems; and how to design content for specific answer contexts like snippets, PAA, and voice.

Part III – Maximizing Visibility in AI and Search

How to win featured snippets, dominate People Also Ask, optimize for voice and local, get retrieved by LLMs, influence training data, and adapt to different channels (web, local, video, books, and AI interfaces).

Part IV – Optimization, Measurement, and the Future

How to monitor performance across platforms, debug issues, manage the lifecycle of your assets, uphold ethical standards, and build an AEO practice that keeps working as the ecosystem shifts.

You'll also find the LLM SEO Toolkit and an AEO maturity model to benchmark where you are.

Every part ends with:

- A brief summary ("what you just accomplished"), and
- A checklist you can use as a working audit for your site or brand.
- Suggested Reading Paths
- Depending on your role and timeline, you do not have to read the book linearly the first time.

If you are an in-house SEO or content lead:

- Read Part I in full to align your mental model of AEO, LLMs, and E-E-A-T.
- Move directly into Part II and apply the pillar/cluster frameworks to one real topic.
- Jump to Part III chapters on snippets, PAA, and LLM retrieval for your priority pages.
- Use the checklists at the end of each part as working documents.

If you are a CMO, VP, or founder:

- Read the Introductions to Parts I–IV and each part's Summary & Checklist.
- Skim the detailed chapters where you need more depth on a specific decision (e.g., entity management, training-data strategy, or ethics).
- Use the Final Checklist and AEO Maturity Model to set expectations and internal KPIs.

If you are an agency strategist or consultant:

- Read Part I and Part III first to get the landscape and visibility levers.
- Then work through Part II with one "flagship" client in mind to design an AEO content roadmap.
- Use Part IV to build a repeatable monitoring and reporting framework across clients.

If you're new to SEO but responsible for content:

- Move slowly through Part I, treating each chapter's key takeaways and checklist as exercises.
- In Part II, focus on the definitions of pillar content, clusters, and structured data.
- Refer back to the glossary and examples whenever a concept feels unfamiliar.

How to Work Through the Material

The best way to use this guide is to anchor it to one real site, brand, or project.

For each part:

- Read a chapter once without stopping.
- Get the full narrative before worrying about details.
- On the second pass, annotate with a specific property in mind.
- Highlight anything that describes a gap you already see in your site.
- Mark any frameworks that could become your next project brief.
- Apply the checklist at the end of the part to that property. Treat the checklist as an audit. Mark items as:
    - Done
    - In progress
    - Not started
- Turn gaps into concrete tasks.
- Move items from "Not started" into your project management system as tickets, sprints, or experiments, with owners and dates.

By the time you reach the final checklist and maturity model, you should have a clear backlog of AEO work broken into actionable pieces, not just a list of abstract ideas.

Using the Checklists and Toolkit

Checklists appear at the end of Parts I–IV, and the LLM SEO Toolkit chapter in Part IV inventories tools by job (research, content creation, schema, monitoring, AI visibility, and so on).

Use the checklists and toolkit:

- **As implementation runbooks**

  When you or your team sit down to work on a specific area—pillar pages, schema, local, voice, or AI visibility—open the relevant checklist and work through it line-by-line.

- **As alignment tools with stakeholders**

  Bring the checklist to meetings with product, engineering, or leadership. Use it to show what "doing AEO properly" actually entails and which steps require their support.

- **As recurring audit templates**

  Revisit the same checklists quarterly or biannually. Mark what has improved, what has regressed, and what new gaps have appeared as platforms change.

- **As training material**

  If you are onboarding new team members, give them a chapter plus its checklist as a focused reading assignment aligned to their responsibilities.

WORKING WITH A TEAM

AEO works best when it is treated as a shared, cross-functional responsibility rather than a solo project.

As you read:

- Tag owner roles in the margins.
- Next to each recommendation, mark whether it belongs to content, SEO, dev/engineering, analytics, or leadership.
- Create a simple AEO roadmap.

After each part, summarise your top three priorities:

- One content change
- One technical/structural change
- One measurement or monitoring change

Use examples as conversation starters.

When a chapter describes a snippet pattern, schema type, or training-data tactic, bring one live example from your own SERPs or AI results to your team and dissect it together.

How to Get the Most Value

You will get the strongest results if you treat this guide as:

- A reference manual - something you return to when you hit a problem in search, AI visibility, or measurement.
- A project blueprint - a way to design sprints, experiments, and roadmaps that move you toward AEO maturity.
- A shared language - a tool for aligning content, SEO, product, and leadership around what "being the answer" actually requires.

You do not need to implement everything at once. Start where impact is highest and friction is lowest. Use the checklists to build early wins, then layer in more advanced tactics as your team and systems mature.

The goal is simple: by the time you finish this book—and work through its exercises—you should have a visible, defensible presence in the answer layer of search and AI, not just another set of pages hoping for clicks.

ANSWER ENGINE
JOURNAL

This book was produced by Answer Engine Journal—an applied research and practitioner-focused publication dedicated to the future of AI visibility, AEO, GEO, and LLM-era search.

Because the field evolves quickly, Answer Engine Journal maintains live, continually updated resources to extend and complement what you're reading here.

The central hub for these materials is:

AnswerEngineJournal.com

Answer Engine Journal operates as an independent publication and knowledge center focused exclusively on the emerging discipline of Answer Engine Optimization. Think of it as a specialized counterpart to traditional search-industry publications—built for the world of AI-powered retrieval.

On the site, you'll find:

- Articles and deep dives that expand the strategic models in this book
- Case studies, analyses, and implementation write-ups
- Tool reviews, schema patterns, and practical playbooks for AEO and LLM SEO
- Updates as platforms like Google, Bing, and major LLMs evolve their answer engines

Most resources are available with no opt-in required.

Acknowledgements

We extend our appreciation to the global community of search engineers, machine learning researchers, data stewards, and information architects whose work continues to shape the landscape of modern retrieval. Their public research, open-source contributions, and long-term dedication to improving how information is structured and surfaced form the foundation on which disciplines like Answer Engine Optimization emerge.

We are equally grateful to the educators, librarians, analysts, and documentation teams who uphold the standards that make knowledge discoverable, trustworthy, and accessible. Their roles—often quiet but essential—ensure that information systems serve not only commercial objectives but the broader interests of users, institutions, and society.

Finally, we acknowledge the researchers and policy voices navigating the intersection of AI, search, governance, and public knowledge. Their efforts to promote responsible innovation and ethical deployment create the conditions for future-facing work in AEO, GEO, and AI visibility to evolve in ways that respect accuracy, context, and the long-term health of the information ecosystem.

Part I:

Foundations of Answer Engine Optimization

Answer Engine Optimization, or AEO, is how you make sure your content is the answer that gets delivered. This isn't about chasing rankings. **It's about being the trusted source, the one both users and AI systems turn to when they need clarity.**

In this section, you'll see what sets AEO apart from traditional Search Engine Optimization (SEO). You'll learn how search engines and large language models (LLMs) interpret intent, why authority matters more than ever, and how to structure your content so it stands out as the definitive answer. Each chapter is built to give you practical steps, not just theory.

You'll notice the approach is direct. We focus on user needs, on building trust, and on making your expertise easy for both humans and machines to recognize. Along the way, you'll find examples, checklists, and frameworks that you can use right away.

You'll learn why AEO is not a trend. It's the new baseline for digital visibility. Search engines like Google, Bing, and platforms like ChatGPT and Perplexity now deliver answers, not just search results. If your content isn't designed for this shift, it gets left behind.

As you move through this part, keep an eye on how your own content stacks up. Are you speaking directly to the questions people ask? Are your answers clear, structured, and easy to trust? If not, you'll find the tools here to close those gaps.

Let's get started. The future of search belongs to those who don't just show up—but who get chosen.

Chapter 1

Introduction to Answer Engine Optimization

What Is Answer Engine Optimization?

Search has changed. People now expect answers, not just lists of links. **Answer Engine Optimization meets this shift by aligning your content with how modern search engines and AI systems deliver responses.** The focus isn't on ranking. It's on becoming the answer that users—and algorithms—trust.

Answer Engine Optimization, or AEO, is about clarity and authority. You're not just trying to appear on page one. You want your content to be the response selected when a question is asked. That means structuring information so machines can understand it, and people can rely on it.

You'll see AEO at work in places like Google's Search Generative Experience, ChatGPT, Perplexity, Bing Copilot, and Alexa. These platforms don't just link—they generate spoken, summarized, or structured answers. Featured snippets, "People Also Ask" boxes, and voice search results are all examples of AEO in action.

Traditional SEO relies on keywords and backlinks. AEO requires comprehension and context. The shift is not just in format, but in function. SEO drives traffic. AEO delivers answers.

To succeed with AEO, your content must be clear, well-structured, and reliable. Write for retrieval systems, not just for search bots. The goal is to become the trusted voice—selected, not just indexed.

You'll notice this in action when a recipe surfaces as a featured snippet, or a financial tip is spoken by a smart assistant. These aren't rankings. They're selections. That's what makes AEO powerful.

THE EVOLUTION OF SEARCH BEHAVIOR

Search habits have transformed. People no longer type short phrases like "weather New York." Now, they ask, "What's the weather like in New York today?" Search engines and AI now prioritize context and intent over keywords.

The timeline is clear. Early search results were just ten blue links. Today, zero-click answers and generative summaries dominate. Google's Search Generative Experience provides AI-driven overviews. Users often get what they need without clicking through.

Click-through rates for traditional search results have dropped. Direct answers now satisfy many queries instantly. You might see a step-by-step guide for a recipe or a concise summary for a historical fact. This shift highlights why being the best answer matters more than being the best-ranked page.

Google SGE and AI overviews pull from trusted sources to create comprehensive responses. For content creators, this means optimizing for retrieval, not just ranking. Structured data and semantic clarity are now essential.

Strategies must now prioritize E-E-A-T: Experience, Expertise, Authoritativeness, and Trustworthiness. Content needs to signal reliability to both users and algorithms. The goal is to become the source that AI chooses to cite.

As AI advances, search behavior will keep changing. Understanding these shifts is the first step to staying ahead.

WHY AEO IS A STRATEGIC IMPERATIVE

The way we search is changing. Google SGE, ChatGPT, and Bing Copilot don't just rank results—they summarize, cite, and speak. Visibility now happens in the answer layer, not just on a list of links.

If your content isn't built for this shift, it disappears from view.

You need to show up in three places:

- In **training data**, so AI can learn from you.
- In **retrieval layers**, so AI can find you.
- In **answer outputs**, so AI can cite you.

That's the new pipeline of trust. If you're not in it, you're out of the conversation.

Citations are the new clicks. When a model references your content in a summary or aloud, it positions you as the authority. These mentions build brand credibility with both users and machines.

Authority comes first. Without it, technical tweaks won't matter. Schema, backlinks, and page speed help, but they don't earn trust alone. What earns it is demonstrating experience, clarity, and subject-matter depth—across formats and platforms.

AEO isn't about gaming a system. It's about showing up where decisions are made—by users and by AI.

Visibility is no longer something you wait for. It's something you engineer.

AEO vs. SEO vs. LLM SEO: What's Actually Evolving

Let's clarify the landscape.

AEO isn't just the next phase of SEO. It's a visible part of a larger shift—one that changes how content is selected, surfaced, and understood.

Traditional SEO was built for engines that crawl, index, and rank. AEO emerged as platforms began delivering direct answers instead of links. Now, we're in a third layer—models.

Large language models (LLMs) don't just rank information. They learn it. They embed it during training, retrieve it during queries, and repackage it in natural language answers. Optimizing for this isn't just good SEO—it's its own discipline.

You might hear it called LLM SEO, GEO (Generative Engine Optimization), or AAEO (Agentic Answer Engine Optimization). The point is this: the content you publish today might not just get ranked—it might get trained on, retrieved from, and spoken aloud.

This introduces two challenges:

- How do I get my content into the datasets models are trained on?
- How do I make sure it gets picked when a tool tries to answer a real question?

AEO is your tactical layer—what makes you eligible to be the spoken or featured answer. LLM SEO is the strategic layer—what makes you the trusted source across tools, agents, and systems.

Here's a quick comparison:

Layer	What You're Optimizing For	How It Works	Your Strategic Goal
Legacy SEO	Web crawlers and ranking	Keywords, backlinks, technical SEO	Show up in search results
AEO	Answer engines (SGE, snippets)	Structured data, clear answers	Be selected as the visible answer
LLM SEO	Large language models	Trainable content, retrieval format	Be cited and remembered by AI

KEY TAKEAWAYS

- AEO is not an SEO update. It's a fundamental reframe that targets how AI systems select and deliver answers, not just how engines rank pages.
- Visibility now happens inside the answer, not around it. If you're not engineered for selection, you're not surfaced.

CHAPTER 2

How Search Engines and LLMs Work

UNDERSTANDING THE ENGINES BEHIND ANSWERS

Search engines and large language models (LLMs) both aim to deliver the best answers, but they approach the task differently. Traditional search engines index and rank web pages. LLMs generate responses by interpreting context and meaning.

To optimize your content for today's answer-driven landscape, you need to understand how both systems operate.

This chapter explains how these engines evolved, how they interpret queries, and where their methods diverge. You'll see why semantic understanding now matters more than keyword matching. You'll also learn how training data and ranking data both play a role in Answer Engine Optimization (AEO).

TRADITIONAL SEARCH ENGINES: THE BASICS

Search engines like Google and Bing still rely on three pillars: crawling, indexing, and ranking.

- **Crawlers discover** new and updated web pages.
- **Indexing stores** these pages for fast retrieval.

Ranking algorithms decide the order in which results appear.

Links, content relevance, and user experience remain critical. Backlinks from reputable sources signal authority. Pages that load quickly and are easy to navigate rank higher. These factors haven't disappeared, even as search evolves.

Structured data has changed the game. Schema markup helps search engines understand your content, not just read it. Featured snippets and rich results reward clarity and direct answers. If your content is well-structured, it's more likely to be surfaced as a direct answer.

User reviews and consistent business information also influence rankings, especially for local searches. Search engines now weigh these signals alongside traditional ranking factors.

The shift from keywords to semantic search marked a turning point. Search engines now look for meaning, not just matches. Voice search and conversational queries accelerated this trend. The goal has moved beyond ranking pages—it's about answering questions.

Tools like Google Search Console and schema validators help you monitor technical health and performance. They're essential for anyone serious about visibility.

Large Language Models (LLMs) and How They Generate Answers

Large Language Models are the "engines" behind most AI interfaces, such as the chat interface for ChatGPT. Some other examples of LLMs are Claude, Google Gemini, and Deepseek.

LLMs like GPT, Gemini, and Claude learn from vast datasets. They break down text into tokens and look for patterns to predict the next word or phrase. This enables them to generate coherent, context-aware responses.

Retrieval-Augmented Generation (RAG) is a major advance. Rather than just using the base knowledge or training data that LLMs come with, the response is informed by "retrieved" data.

In the context of the ChatGPT chat interface, this would happen after you upload a document or it searches for (and crawls) a page on the internet. Your response would have data from the uploaded document or the page that was retrieved.

RAG combines the model's internal knowledge with real-time data fetched from external sources. This keeps answers current and accurate.

LLMs prioritize authoritative sources. They scan trusted websites, extract relevant information, and cite references. Consistency and clarity matter. Structured data and schema markup make it easier for models to understand and retrieve your content.

In terms of how you or your business come up in web-based RAG output, brand consistency is key. Make sure your business or product is represented uniformly across platforms like Google Business Profile and Wikipedia. This helps LLMs connect the dots.

Monitoring LLM outputs is still evolving. Manual checks and emerging tools can help you track how often your content is cited or referenced. The goal is to be included—and trusted—as a source.

Semantic Comprehension vs. Keyword Matching

AI systems now interpret meaning through context, not just keywords.

Traditional search rewarded exact phrase matches. Modern systems now use LLMs and contextual embeddings to understand intent.

Clear, natural language matters. Ambiguous phrasing confuses AI, while precise wording improves answer quality. For example, instead of just targeting "best dog food for allergies," address the underlying need. Explain hypoallergenic options, compare brands, and cite expert opinions.

Structured data also bridges the gap between human and machine understanding. Schema markup clarifies relationships between entities, helping AI categorize information accurately. We will go over schema and metadata in detail later, but for now we will broadly categorize it as, "data about data".

Training Data and Ranking Data: The Two Fronts of AEO

AEO operates on two parallel tracks:

- **Training Data:** This is what AI models learn from. Structured, authoritative sources—like books, PDFs, and well-organized web content—teach AI how to answer questions. Schema markup is essential here.
- **Ranking Data:** This determines what gets retrieved in real time. Freshness and authority matter. Schema markup and backlinks from trusted sites reinforce reliability.

Tools like IndexNow and Google-Extended help search engines and AI models find your new content quickly. These tools bridge the gap between being learned and being found.

Success in AEO comes from balancing both fronts. Training data builds long-term knowledge. Ranking data ensures your content is accessible when it matters.

Key Takeaways

- Search engines and LLMs follow different paths but share the goal of delivering trusted answers.
- Semantic clarity now outweighs keyword repetition. AI reads for meaning, not just matching.

CHAPTER 3

Information Authority in the Age of AI

WHY AUTHORITY MATTERS MORE THAN EVER

Trust is the foundation of modern search. As AI-driven answers become the default, credibility is no longer optional—it's essential. Search engines and large language models (LLMs) now use authority signals to decide which sources to surface and cite. **If your content doesn't demonstrate expertise, experience, and reliability, it won't be chosen.**

You'll notice this shift in how Google and other platforms evaluate content. Their E-E-A-T framework—Expertise, Experience, Authoritativeness, and Trustworthiness—guides what gets shown. The same standards now influence AI-generated answers. Building authority is no longer just a way to boost rankings. It's the key to being the answer.

UNDERSTANDING E-E-A-T

Google's E-E-A-T framework shapes visibility in both search and AI. Each element plays a distinct role:

- **Expertise**: Content must be created by someone with proven knowledge. Credentials, professional experience, or a deep track record in the field all count. Author bios should highlight these qualifications.
- **Experience**: Firsthand accounts and practical insights matter. For product reviews or how-to guides, personal narratives and specific examples win trust.
- **Authoritativeness**: Recognition by others is critical. Backlinks from respected sites, citations in industry publications, and mentions in research all contribute.

Internal linking to your own high-quality content also helps.

- **Trustworthiness**: Accuracy and transparency are non-negotiable. Avoid exaggerated claims or manipulative tactics. User reviews and testimonials, when genuine, reinforce trust.

Consistency across these signals is vital. If your business details or author credentials vary from page to page, algorithms may hesitate to trust you.

BUILDING AND SIGNALING KNOWLEDGE AUTHORITY

Publishing in respected venues—whether it's a technical guide, a research archive, or a well-known blog—amplifies your authority.

Certain formats carry more weight. Books with ISBNs, open-access PDFs, and white papers all signal deep expertise. These assets often become reference points for both users and AI systems.

Linking strategies matter, too. Backlinks from reputable sources validate your expertise. Internal links create a web of credibility within your own site. Schema markup, such as author, acceptedAnswer, or Person, makes your authority machine-readable.

Focus on niche dominance. Become the definitive source for your topic. Comprehensive, regularly updated content ensures you stay relevant. Tools like Google Search Console can help you spot gaps and opportunities.

AVOIDING RED FLAGS THAT UNDERMINE AUTHORITY

Avoid these red flags to prevent getting penalized by search and answer engines.

Thin content is a warning sign. Shallow articles, generic summaries, or copied material erode trust. Search engines and LLMs look for depth and originality. Reviews and recommendations should show real research, not just surface-level opinions.

Technical errors can also harm your reputation. Broken links, soft 404s (error pages that don't return the correct status code), and inconsistent metadata confuse crawlers and users. Malware, deceptive downloads, or manipulative practices trigger severe penalties.

Metadata must be precise. Conflicting schema or vague descriptions mislead both people and machines. Use standardized values from schema.org to ensure clarity.

User experience counts. Too many ads, cluttered layouts, or generic 404 pages frustrate visitors and invite demotion. Prioritize helpful navigation and clean design.

Content farms and mass-produced articles without clear authorship lack credibility. Attribution and original research build trust. Regular audits catch issues before they damage your standing.

KEY TAKEAWAYS

- Trust is now a primary ranking signal. Both search engines and LLMs favor content that demonstrates expertise, accuracy, and transparency.
- E-E-A-T is your compliance framework. Author bios, citations, structured data, and genuine user trust signals all contribute to your authority.

CHAPTER 4

Framing AEO Within Broader Search Intent

WHY SEARCH INTENT MATTERS

Every search begins with intent. Whether someone wants an answer, a product, or a recommendation, their goal shapes the way they search—and the way engines respond. **If your content doesn't match user intent, it won't get chosen, no matter how well it's written.**

Search engines and large language models (LLMs) now focus on intent as much as keywords. They look for content that fits the user's purpose, not just their phrasing. When you align your work with intent, you increase your chances of becoming the answer.

TYPES OF SEARCH INTENT AND THEIR ROLE IN AEO

There are four primary types of search intent:

- **Informational**: The user wants to learn something. Examples include "how does solar power work" or "tips for remote teams." AEO excels here by providing clear, complete answers. Pillar pages and detailed guides are especially effective.
- **Commercial**: The user is researching solutions or wants to learn more about a specific product. Searches will be for detailed information and will often include terms like "best" or "cheapest." Comparison pages work well for this.
- **Transactional**: The user is ready to act—buy, sign up, or download. Queries like "buy running shoes" or "sign up for newsletter" need direct, actionable content. Product pages and clear calls-to-action work best.

- **Navigational**: The user wants to reach a specific site or page. Think "Twitter login" or "Harvard admissions." For AEO, make sure your brand's navigational queries resolve cleanly and quickly.

Matching your content to intent means going beyond keywords. Use structured data, like FAQ or HowTo markup, to help AI understand the context. Tools like SEMrush, Ahrefs, or Google's "People Also Ask" can help you discover the real questions users have.

How LLMs Evaluate and Select Content

LLMs don't just look for keywords. They prioritize:

- **Authority**: Is the content trustworthy and well-sourced?
- **Clarity**: Is the answer direct and easy to understand?
- **Relevance**: Does the content fit the question's context?
- **Consistency**: Are facts, names, and details uniform across your site and profiles?

Verbose, hedging, or ambiguous language gets filtered out. LLMs prefer direct, factual statements. If your answer is opinion-based, label it clearly. When possible, back up claims with evidence.

Training data shapes what LLMs know. The more your content is cited or referenced by trusted sources, the more likely it is to be embedded in future AI responses. Retrieval-augmented generation adds another layer: LLMs pull in fresh data from trusted sites to refine their answers. If your content is well-structured and authoritative, it's more likely to be selected.

Your Content's Journey to Becoming an Answer

Content takes two main paths to reach users:

- **Traditional Search**: Crawling, indexing, and ranking. Here, you compete to be one of many results.
- **Answer Engines**: Training, retrieval, and citation. Here, you compete to be the single answer.

The difference is significant. Traditional search rewards relevance and authority. Answer engines demand trust, clarity, and structure. Weak or vague content gets ignored. Strong, well-structured content rises to the top and gets cited.

Structure is key. Use clear headings, logical flow, and semantic markup. Pillar content anchors your expertise, while supporting content (like FAQs or blog posts) fills in gaps and reinforces your authority.

E-E-A-T—Experience, Expertise, Authoritativeness, Trustworthiness—remains the deciding factor. Content that demonstrates all four is more likely to be chosen as the answer.

Key Takeaways

- Search intent is the foundation of modern visibility. Align your content with what users want, not just what they type.
- LLMs and search engines evaluate for clarity, authority, and relevance. Direct, well-structured answers outperform clever or ambiguous phrasing.

CHAPTER 5

Summary and Action Steps

LOOKING BACK: WHAT YOU'VE LEARNED

If you've made it this far, you're ahead of the curve. You now understand why Answer Engine Optimization (AEO) is more than a buzzword. It's a response to a real shift in how people search—and how answers get delivered. Search isn't just about links anymore. It's about being the answer, the one both users and AI trust.

Throughout Part I, you've seen how AEO diverges from traditional SEO. You've learned that search engines still crawl, index, and rank, but large language models (LLMs) retrieve and generate answers based on patterns and context. To show up in these new systems, your content must speak their language: clarity, authority, structure, and semantic intent.

You've also explored E-E-A-T: Experience, Expertise, Authoritativeness, and Trustworthiness. These aren't just checkboxes. They're signals that both Google and AI models use to decide if your content is worth citing or ignoring. You've learned how to make your content readable by machines, relevant to humans, and structurally sound for training and retrieval.

The heart of AEO is shifting from chasing rankings to earning relevance. It's about being cited, not just clicked. Whether you write guides, publish research, or manage a brand, your new goal is to be the most helpful, reliable, and complete source—for people and for the machines that answer on their behalf.

AEO Foundations — Checklist

Understanding the Landscape

- ❒ I can clearly explain what Answer Engine Optimization (AEO) is and how it differs from traditional SEO.
- ❒ I understand why AEO is essential in the era of AI search, zero-click results, and conversational interfaces.
- ❒ I know what qualifies as an "answer engine" (e.g., Google SGE, ChatGPT, Perplexity, Bing Copilot, Alexa).

Search Systems & LLMs

- ❒ I understand how traditional search engines crawl, index, and rank content.
- ❒ I can describe how large language models (LLMs) retrieve and generate answers using both training data and real-time retrieval.
- ❒ I grasp the basics of Retrieval-Augmented Generation (RAG) and why formatting for machine comprehension matters.

Semantic Optimization

- ❒ I understand the difference between keyword matching and semantic comprehension—and why clarity and structure now outweigh keyword repetition.
- ❒ I know how content phrasing and structure can determine whether it gets selected as an answer by AI.
- ❒ I've considered how to format my content for predictable, citation-ready output.

Authority & Trust Signals

- ❒ I understand the E-E-A-T framework (Experience, Expertise, Authoritativeness, Trustworthiness) and how it applies to both search engines and AI content selection.
- ❒ I've identified gaps in my content or web presence that could weaken my authority or trustworthiness.
- ❒ I know how publishing formats like PDFs, books, and structured knowledge (e.g., ISBNs) can strengthen perceived expertise.

Strategic Mindset Shift

- ❐ I recognize that my primary goal is not just traffic, but visibility in AI systems and long-term knowledge authority.
- ❐ I'm thinking beyond single-page rankings—toward content that can be cited, retrieved, and repurposed by AI.
- ❐ I've started to view content as an information asset that earns trust over time, not just a temporary traffic driver.

You've now got the foundation. In the next section, we'll show you exactly how to build on it — starting with the creation of pillar content that not only ranks, but earns a permanent place in the answer economy.

PART II

Strategic AEO Content Development

Modern search rewards content that answers, not just content that ranks. You've seen how foundational principles like clarity, authority, and structure set the stage for visibility in answer engines and AI-driven platforms. Now, it's time to put those principles into practice.

In this section, you'll move from understanding to execution. You'll learn how to create pillar content that anchors your expertise, structure information for both human readers and machines, and build supporting assets that reinforce your authority across topics. The goal is to make your content not just discoverable, but indispensable—chosen by users and cited by AI systems.

We'll walk through frameworks for building comprehensive resources, using structured data to enhance retrieval, and optimizing for featured snippets, voice, and conversational queries. You'll see how to map internal content ecosystems, leverage technical tools, and distribute your work so it's included in both real-time retrieval and long-term AI training sets.

Let's get started. The next phase is about building assets that last, scale, and earn trust—no matter how search evolves.

CHAPTER 6

Creating and Scaling Pillar Content for AEO

WHAT IS PILLAR CONTENT IN THE AEO CONTEXT

Pillar content is the backbone of your AEO strategy. It's a comprehensive, authoritative resource that covers a broad topic in depth, answers core questions, and serves as the hub for related subtopics.

When you build pillar content, you're aiming to become the definitive source—one that both users and AI systems cite and return to over time.

Traditional SEO content often targets narrow keywords or fleeting trends. Pillar content takes a different approach. It's built for longevity and depth. It anticipates the questions users will have, addresses them directly, and links out to supporting cluster content that dives deeper into specific angles. The goal is not just to rank, but to be the answer that gets surfaced, cited, and trusted.

Authority is central. Pillar content demonstrates your expertise, experience, and trustworthiness. It's regularly updated to stay relevant. When done well, it becomes the anchor for your entire content ecosystem. Supporting articles, FAQs, and news updates link back to it, reinforcing its authority and helping both search engines and large language models (LLMs) recognize it as the primary resource on the topic.

Structure matters. Clear headings, semantic organization, and internal links guide readers and crawlers alike. Metadata and schema markup clarify the page's purpose. When you focus on building a strong pillar, you create a foundation that supports all your related content—and your visibility across answer engines.

Anatomy of High-Performing Pillar Pages

A high-performing pillar page balances depth with clarity. It guides readers through complex topics without overwhelming them. The structure is intentional, designed to answer every major question while making navigation easy.

A table of contents at the top lets users jump to sections that matter most. Logical headings break the content into manageable pieces. Each section serves as a standalone answer, reinforcing the overall authority of the page.

Summaries and embedded answers distill key points. Collapsible FAQs address common questions without clutter. Infographics and tables visualize data for quick understanding. Citations back up claims, building trust with both users and AI.

Scannability is critical. Use bullet points, bolded phrases, and white space to keep the reader engaged. Semantic clarity ensures that both humans and machines understand the content's intent and structure.

Internal linking strengthens the page's role as a hub. **Schema markup**, such as Article, FAQPage, or HowTo, signals the type of content you're providing. Fast indexing and regular updates keep the page current and authoritative.

Every element—from navigation to citations—works together. The result is a resource that educates, earns trust, and stands as the go-to answer for its topic.

Creating Content Designed to Be Cited by LLMs

If you want your content to show up in AI-generated answers, you need to write with citation in mind. LLMs don't just retrieve—they predict. They surface clear, unambiguous, and semantically rich information that stands alone, even when quoted out of context.

Lead with the answer. Avoid long intros or filler. Each paragraph should make sense on its own, since AI often cites fragments or summaries. Use direct, factual language. Hedging or vague statements get filtered out.

Test your content with AI tools. Prompt ChatGPT or Claude to summarize your paragraphs. If the summary isn't accurate or clear, revise your original copy. Semantic scoring platforms like SurferSEO or Clearscope can help you identify gaps in coverage or clarity.

Don't be afraid to use AI for drafting or outlining. These tools can help surface examples, structure sections, or generate FAQs. But always review for accuracy, tone, and authority. LLMs are more likely to cite content that reads like a well-edited reference than a marketing pitch.

Beyond summaries, use AI as a strategic editor. Ask it to find gaps, missing proofs, and better angles—not just to rewrite paragraphs. We will later explain how search engines and answer engines handle related concepts and topical depth. For now, here are some prompts you can use to improve both coverage and usefulness:

- "You've just summarized this article. What important questions would a **[job title]** still have after reading it?"
- "Given my goal of attracting **[specific audience]**, what subtopics or sections are missing from this guide?"
- "Based on this article and my positioning **[I help X achieve Y through Z]**, what would be three high-leverage collaboration ideas or podcast topics that reinforce my authority?"
- "If you were an editor at an industry publication for **[audience]**, what would you ask me to clarify or prove with data before publishing this?"

Remember, you're writing for both readers and retrievers. The clearer and more structured your content, the more likely it is to be surfaced and cited by answer engines.

Metadata, Titles, and Semantic Clarity

Clarity is your top priority. Machines and humans both prefer straightforward titles and metadata. Avoid cleverness that could confuse either audience. Titles should communicate the topic instantly. Metadata should reinforce the page's purpose.

AEO titles focus on precision. For example, "How to Optimize for Voice Search in 2026" is clear and direct. Avoid ambiguous or playful headlines that don't translate well for machines.

Align metadata with content intent. Use schema markup to label your information. This helps search engines and LLMs parse your page correctly. Structured data turns descriptions into machine-readable signals, increasing your chances of being selected as an answer.

When titles and metadata reflect the content accurately, you bridge the gap between discovery and understanding. This alignment leads to better performance in both search and answer engines.

Generative vs. Synthetic Content

To make great content with AI tools, you need to understand the difference between **generative** and **synthetic** content.

Generative content starts from something real – a recorded call, a case study, analytics, product documentation, or a live demo. AI then helps you clean, expand, structure, and repurpose that material.

Synthetic content starts from nothing – a vague prompt like "write an article about back pain for chiropractors in Denver" with no underlying data, story, or real experience.

Any AI tool you use will deliver better output when you ensure high-quality input. As people say in data work: "garbage in, garbage out."

Good data sources for generative workflows include:

- Recorded Zoom or phone calls with clients
- Internal SOPs, implementation notes, and debug logs
- Before-and-after results from campaigns, treatments, or product changes
- Live demos, webinars, or conference talks

From there, AI can:

- Transcribe and clean the language
- Extract key steps, timelines, and metrics
- Turn real events into case studies, FAQs, and how-to articles

For AEO content, always work generatively, not synthetically. Answer engines increasingly favor content grounded in real-world experience, specific outcomes, and verifiable details. *Especially* your pillar content.

Purely synthetic articles may rank briefly, but they usually fail E-E-A-T checks and user tests. They read generic because they are generic.

AI Tools to Scale Your AEO Content

You don't need a massive team to scale pillar content. AI tools can help you move faster and smarter. However, you should always start from high quality source material based in reality.

To discover trending questions and gaps in existing coverage, use platforms like Frase, BuzzSumo, or MarketMuse. These tools reveal what users are asking—and what's missing from current search results.

Drafting and outlining become easier with platforms like Jasper, Writesonic, or Claude. They can generate draft sections, repurpose content into FAQs, or create outlines based on your headings. You remain in control of quality, voice, and structure.

Semantic optimization tools like SurferSEO or Clearscope analyze your drafts for depth and coverage. They highlight missing entities or subtopics, ensuring your content meets the expectations of both users and AI.

As a final check, simulate retrieval behavior by prompting models like ChatGPT with the entire draft. Ask, "What are the key points from this article?" If the model struggles, misses major arguments, or latches onto minor details, refine your copy and structure. The goal is to make your content easy for AI to summarize and cite at both the paragraph and page level.

AI accelerates your workflow, but your judgment ensures the result is authoritative and trustworthy. Use technology as a partner, not a replacement.

Key Takeaways

- Pillar content anchors your AEO strategy. It's comprehensive, authoritative, and built to last.
- Structure, clarity, and internal linking make pillar pages the go-to resource for both users and AI.

CHAPTER 7

Structured Data and Technical Optimization

WHY STRUCTURE MATTERS IN AEO

Structured data is the bridge between your expertise and the systems that surface it. It tells search engines and large language models (LLMs) what your page means, not just what it says.

Without clear structure, even the most helpful content can be missed or misunderstood by machines. **If you want to be the answer, you need to speak their language.**

Schema.org provides the vocabulary. JSON-LD is the preferred format for embedding this vocabulary into your site's HTML (code). It's clean, flexible, and easy to update. Microdata works too, but JSON-LD keeps your code organized and your markup separate from visible content.

When you use structured data, you unlock rich results. These might include review stars, event dates, or product details in search. For AEO, the impact goes further. LLMs use structured data to interpret, summarize, and sometimes even quote your content in answers. A well-marked page can become the foundation for a spoken result or a featured snippet.

Precision is critical. Missed fields or syntax errors can weaken your signals. Tools like Google Search Console and schema validators help you catch problems before they cost you visibility. Properly applied, structured data turns your content into clear, actionable information for both users and machines.

Structure isn't optional. It's the foundation of visibility in a world where machines curate answers. When your content is easy for them to read, it's more likely to be chosen.

CRITICAL SCHEMA TYPES FOR AEO

Not all schema types matter equally for AEO. Focus your efforts on those that directly support answer extraction and authority signals:

- **FAQPage**: Marks up question-and-answer content, making it eligible for rich results and voice responses.
- **Article**: Defines written content, clarifies authorship, and signals publication dates.
- **Person** and **Organization**: Establish identity and connect your content to real-world entities.
- **HowTo**: Breaks down processes into clear steps, ideal for instructional content.
- **LocalBusiness**: Highlights location, hours, and services for local queries.
- **Speakable**: Identifies content suitable for voice assistants.
- **Review** and **Event**: Structure feedback and event details for easy retrieval.
- **QAPage**: Organizes question-and-answer formats for forums or community content.

Implementation matters. JSON-LD is the standard for embedding schema. Use it to keep your markup clear and separate from your main content. For videos, include attributes like `uploadDate` and `interactionCount` to help search engines and LLMs understand relevance and popularity.

Complement schema with technical tags. Robots meta tags manage crawling, while hreflang tags handle multilingual content. Together, they ensure your structured data works as intended.

The right schemas make your content machine-friendly. When search engines and LLMs can interpret your pages easily, they're more likely to surface your answers.

Embedding AEO-Friendly Media

Media can boost your authority, but only if machines can interpret it. Videos, images, and audio need clear, structured support to be surfaced in AI answers.

Start with the video upload. Use descriptive titles that match search intent. Provide full transcripts with speaker markers and timestamps. This makes spoken content machine-readable and accessible. Backlink your videos contextually within related articles to reinforce authority.

Video schema is essential. Mark up your videos with attributes like uploadDate, duration, and description. This helps search engines and LLMs feature your content in answers and summaries.

Images need semantic alt-text. Avoid generic file names. Instead, use specific, descriptive labels that clarify what the image shows. Structured data for images, such as thumbnailUrl and uploadDate, further enhances discoverability.

Monitor performance in Google Search Console. Fix flagged issues with structured data or missing metadata. Trends tools can help you identify new topics or formats to keep your media relevant.

AEO-friendly media isn't just visible. It's actionable. When machines understand your videos and images, they're more likely to share them in answers.

Accessibility and Crawlability

If search engines can't access your content, it doesn't exist. Technical optimization ensures your site is discoverable, indexable, and accessible to both users and machines.

Use robots.txt and noindex tags wisely. Don't block critical resources like CSS or JavaScript.

Canonical tags help avoid duplicate content issues. Canonicals guide crawlers to the authoritative version of each page and also prevent page authority cannibalization.

Mobile-first design is no longer optional. Google prioritizes mobile-friendly pages, and slow load times can hurt both user experience and crawl efficiency. Regularly test your site on different devices to ensure speed and usability.

Interlinking is key. Broken links disrupt both crawlers and users. Internal links distribute authority and help machines understand which pages matter most. Use tools to audit for 404 errors and fix them promptly.

Structured data supports indexing. Use schema like `ItemList` to clarify relationships between pieces of content. Competitor analysis can reveal gaps in your technical setup and inspire improvements.

Accessibility goes beyond code. Use semantic HTML, alt-text, and keyboard navigation to ensure everyone can use your site.

Using alt-text for its original designed purpose, to describe images for the visually impaired, is also a great way to make sure your media is SEO and AEO-friendly. This benefits both humans and machines, increasing your reach and trust.

Crawlability is an ongoing commitment. When your technical foundations are solid, your content is positioned to thrive.

KEY TAKEAWAYS

- Structured data is essential for AI visibility. Schema markup makes your content understandable to machines, increasing its chances of being retrieved, cited, or featured in answers.
- Technical optimization directly impacts performance. Crawlability, mobile readiness, and loading speed determine whether your content gets seen at all.

CHAPTER 8

Supporting Content Strategy and Internal Ecosystems

WHY SUPPORTING CONTENT MATTERS IN AEO

Pillar content anchors your authority, but it doesn't stand alone. A robust Answer Engine Optimization (AEO) strategy thrives on an ecosystem of supporting content—blogs, FAQs, news updates, and resource pages—that surround and reinforce your main topics. **These assets answer niche questions, address emerging trends, and drive engagement from both users and AI systems.**

Supporting content builds context. It fills the gaps that pillar pages can't cover in depth. When you interlink these assets, you help search engines and large language models (LLMs) map your expertise across a topic. The result is a network of authority that boosts your visibility and trust.

BUILDING A TOPIC CLUSTER MODEL FOR AEO

A **topic cluster** is a group of related content built around a central pillar page. Each supporting piece explores a subtopic, question, or scenario in detail, then links back to the pillar. This structure signals to search engines and LLMs that you cover the subject comprehensively.

Mapping clusters starts with research. Use tools like SEMrush, Ahrefs, or Google's People Also Ask to uncover the questions real users are asking. Identify gaps in your coverage and prioritize subtopics that align with your audience's needs.

Internal linking is key. Every supporting article should connect to the pillar and, where relevant, to other cluster pieces. This web of links guides crawlers and users through your expertise. Schema markup, such as BreadcrumbList or FAQPage, clarify relationships.

Semantic mapping expands your reach. Cover variations of core questions, related entities, and emerging trends. Passage ranking—the process by which search engines evaluate individual sections—rewards depth and clarity. Structure your content to highlight key insights, making them easy for AI to retrieve.

A well-built cluster ecosystem ensures your pillar page stands as the definitive answer, supported by a network of relevant, up-to-date resources.

Keyword Research Tools and How to Use Them

Effective supporting content starts with understanding what people search for. Keyword research tools reveal the language, questions, and intent behind user queries.

- **SEMrush** and **Ahrefs** offer comprehensive analysis, including search volume, difficulty, and competitor insights.
- **Google Keyword Planner** provides direct data from Google, useful for gauging demand.
- **People Also Ask** and autocomplete features surface long-tail and conversational queries.
- **Local SEO tools** like BrightLocal help you target region-specific questions.

Cross-reference data from multiple tools for a balanced view. Focus on intent, not just volume. Related searches and PAA results help you identify subtopics and question clusters that deserve their own supporting articles.

Competitive analysis reveals gaps. Study what your rivals cover—and what they miss. Use this insight to position your supporting content where it can fill a need or answer a question better than anyone else.

Blog, News, and FAQ Content That Reinforces Pillars

Blogs and news updates keep your ecosystem dynamic. They address timely questions, industry developments, and niche scenarios. Each piece should link back to the relevant pillar page, reinforcing your authority and guiding both users and AI through your content network.

Question-based titles mirror how people search and how AI systems interpret queries. Use schema markup like FAQPage or HowTo to structure answers for easy retrieval. Each question becomes a potential entry point, driving engagement and increasing your chances of being cited.

E-E-A-T—Experience, Expertise, Authoritativeness, Trustworthiness—applies at every level. Highlight author credentials, cite reputable sources, and keep your content accurate and up to date. Tools like AIOSEO can automate schema generation and optimize titles for both users and machines.

Supporting content is not secondary—it's strategic. When blogs, FAQs, and news reinforce your pillars, they create a cohesive system that adapts to trends and changing user needs.

Key Takeaways

- Interlinked content ecosystems drive authority. Topic clusters built around pillar pages send clear signals to search engines and LLMs that you're a topical leader.
- Supporting content fills strategic gaps. Blogs, FAQs, and timely news expand your coverage and strengthen your domain's authority when properly linked back to pillar content.

CHAPTER 9

Creating Content Designed for Specific Answer Contexts

WHY CONTEXT MATTERS IN ANSWER ENGINE OPTIMIZATION

Not all answers are delivered the same way. Featured snippets, "People Also Ask" (PAA) boxes, and voice search each require tailored content. If you want your work to appear where users are looking—and where AI is listening—you need to design your answers for these specific contexts. **This chapter walks you through how to structure, phrase, and format your content so it's ready to be selected, cited, or spoken.**

Context shapes visibility. A paragraph that works for a featured snippet might not be ideal for voice search. A list that wins a PAA box could be ignored in a conversational interface. Your goal is to anticipate these differences and build flexibility into your content. The more adaptable your answers, the wider your reach across AI-driven search.

FEATURED SNIPPETS AND ZERO-CLICK RESULTS

Featured snippets on Google are prime real estate. These short, direct answers appear above the first organic search result—often called "position zero." They also feed into generative AI summaries, voice assistants, and chat interfaces. To win a snippet, your content needs to be both clear and structured.

Start with the answer. Place the core response immediately after a relevant heading. Don't bury it in a long introduction. For definition queries, use a single, direct sentence. For process or comparison queries, use numbered steps, bullet points, or tables.

Keep answers concise. Aim for 40 to 60 words per snippet-friendly response. Dense or rambling content gets skipped. Use schema markup—FAQ, QAPage, or HowTo—where it fits. While schema doesn't guarantee a snippet, it helps machines interpret your intent and structure.

Snippets reward clarity, not cleverness. Write as if you expect your answer to be quoted directly. If you want to encourage further engagement, offer a partial answer in the snippet and invite users to read on for more detail.

Winning a snippet can reduce click-through rates, especially on mobile. But it significantly boosts brand exposure and authority. In the answer economy, being the answer is often more valuable than being a click.

We will explain featured snippets in further detail in chapter 12.

People Also Ask (PAA) Targeting

The "People Also Ask" box on Google is dynamic. When someone clicks a question, more related questions appear. This creates a branching tree of intent, giving you multiple chances to surface as the best answer.

To target PAA, use the question itself as a heading—formatted as an H2 or H3. Follow with a direct, self-contained answer. Keep these responses tight, usually 40–50 words. Don't sell or ramble. Just answer.

Cover semantically similar variations. If "What is schema markup?" is common, also address "How does schema markup work?" or "Why use schema markup?" These near-duplicate questions expand your reach within the same topic cluster.

Embed questions natively in your content, not just in FAQ sections at the bottom. Google favors answers that appear in context. Use FAQ schema when grouping questions, but prioritize natural, in-content answers.

To find high-value questions, expand the PAA box manually or use tools like AlsoAsked, SEMrush, or Ahrefs. These reveal deeper question maps and help you identify subtopics to address.

PAA is about mapping the knowledge graph. The more questions you answer, the more connections you create—both for users and for AI.

We will explain PAA in further detail in chapter 13.

Structuring Content for Voice Search

Voice search is a winner-take-all environment. There's no list of results—just one answer spoken aloud. To be chosen, your content must be clear, brief, and conversational.

Write for the ear. Use short sentences and plain language. If a sentence feels awkward when spoken, rewrite it. Most voice answers are 40–50 words. Lead with the answer, then provide supporting details if needed.

Use contractions and a neutral, helpful tone. Avoid jargon and complex phrasing. If your answer sounds stiff when read aloud, it won't be chosen. Read your content out loud during editing. If you stumble, revise.

Local and transactional queries are especially important for voice. Make sure your Google Business Profile is up to date and consistent. Use LocalBusiness schema and include clear calls to action, such as "Call now" or "Order online." Customer reviews with relevant keywords can also boost your chances of being selected.

Test your answers on real devices. Use Google Assistant, Siri, or Alexa to ask the questions you want to rank for. If your content isn't being spoken, review your structure, schema, and phrasing.

We will further explain how to rank for voice search in chapter 14.

Adapting to Conversational and Follow-Up Queries

Conversational AI and chat interfaces often present answers in a dialogue. They may ask follow-up questions or pull from multiple sources to build a complete response. Your content needs to anticipate these interactions.

Structure your content so each section can stand alone. Use clear headings, direct answers, and logical flow. Anticipate follow-up questions and address them within the same page. For example, after answering "What is semantic SEO?" include a section on "How does semantic SEO work?" or "What are the benefits of semantic SEO?"

Use schema markup to clarify relationships between questions and answers. This helps AI systems understand the context and pull the right information for multi-part queries.

Conversational interfaces reward content that's easy to break apart and recombine. The more modular your answers, the more likely they are to be used in chat, voice, and generative summaries.

Shaping Content for E-E-A-T

Not every answer is a step-by-step how-to. Sometimes answer engines pull in who you are—your bio, your firm's positioning, or your profile text—as part of the response. To make that kind of expertise machine-readable, give every author and brand a clear, repeatable positioning statement.

You can use a simple X–Y–Z Positioning Line:

> "I help **X** achieve **Y** through **Z**."

Where:

- **X** = the specific audience you serve
- **Y** = the concrete outcome you create
- **Z** = your method, framework, or specialty

For example:

- "I help multi-location dental practices grow new patient bookings through data-driven local SEO and patient education content."
- "I help B2B SaaS founders get sales-ready demos through technical video walkthroughs and answer-engine-ready documentation."

Use the same X–Y–Z line in:

- Your website bio, About page, author page
- LinkedIn headline and summary
- Google Business Profile description and other major profiles

This gives search engines and answer engines a clear, repeated semantic anchor for who you are, who you serve, and what you do. When you also encode that same positioning in structured data (Person, Organization, LocalBusiness), it becomes easier for models to connect your content to the right queries and entities.

Key Takeaways

- Context shapes answer design. Featured snippets, PAA boxes, and voice search each require specific structures and phrasing.
- Lead with the answer. Place direct, concise responses immediately after relevant headings.

CHAPTER 10

Asset Distribution and Knowledge Authority Building

MAKING YOUR CONTENT DISCOVERABLE BY LLMS

Visibility starts long before someone asks a question, it starts with being discoverable.

For your content to be cited by large language models (LLMs) and answer engines, it must be accessible, well-structured, and present in the right places.

Open access is essential. Publish key resources as PDFs on platforms like Internet Archive or Open Library. These repositories are routinely crawled by AI systems, increasing your odds of inclusion.

Technical and research content benefits from distribution on specialized platforms. GitHub, arXiv, and SSRN are trusted sources for LLMs. When you share your work there, use clear metadata, semantic titles, and schema markup to reinforce your authority.

Citations matter. When respected sources reference your work, it signals value to both search engines and AI. Focus on accuracy and originality—fact-check every claim. Errors or inconsistencies can get your content filtered out during training or retrieval.

Monitor accessibility. Use tools like Google Search Console to spot indexing issues. Check for accidental noindex tags or blocked resources. If LLMs can't crawl your content, it won't be included in their knowledge base.

Avoid manipulative tactics. Keyword stuffing, hidden text, and low-quality backlinks can trigger penalties or exclusion. Instead, prioritize quality, relevance, and ethical optimization. The goal is to be a trusted, retrievable source—not just a visible one.

Distribution is strategic. Publish where AI systems look, use formats they can parse, and maintain your content for long-term accessibility. When you do, your work becomes part of the information ecosystem that powers AI-driven answers.

Entity Optimization and Semantic Authority

Entities are the building blocks of modern search. They represent people, organizations, products, and concepts—each with a unique identity in the knowledge graph.

Optimizing for entities means making your presence clear, consistent, and authoritative across the web.

Start with schema markup. Use `@type` and `@context` to define your entity—Person, Organization, Product, or Event. Include attributes like `affiliation`, `sameAs`, and `knowsAbout` to connect your entity to related topics and profiles. This helps AI distinguish you from others with similar names or roles.

Consistency is key. Use the same name, bio, and branding across your website, social media, and third-party profiles. Link your main entity page to profiles on LinkedIn, Crunchbase, Wikidata, and other authoritative directories. Internal links reinforce these connections and clarify your expertise to both users and machines.

Co-occurrence matters. Mention your name, brand, or product alongside relevant keywords and concepts. This strengthens your association with those topics in the eyes of AI and search engines. For example, a technology consultant should appear consistently with terms like "AI optimization," "machine learning," and "digital strategy."

Entity-level E-E-A-T (Experience, Expertise, Authoritativeness, Trustworthiness) applies beyond pages. Highlight credentials, awards, and verified profiles wherever possible. The more signals you provide, the easier it is for AI to recognize and trust your entity.

To move from being a mention to becoming a recognized node, pursue corroboration. Secure listings in directories, contribute to Wikidata, and publish on structured platforms like GitHub or SSRN. These third-party signals build your semantic resume and increase your chances of being surfaced in answer engines.

ESTABLISHING EXTERNAL AUTHORITY

External authority amplifies your reach. When other reputable sites and platforms reference your work, it validates your expertise and expands your influence.

Start by pursuing citations in respected media, industry blogs, and academic publications. Guest posts, interviews, and collaborations can introduce your brand to new audiences and reinforce your authority. Use schema like VirtualLocation or Place for events, and register books with ISBNs to ensure inclusion in Google Books and Amazon.

Monitor your online presence. Tools like BrightLocal help you find and fix outdated or incorrect listings. Report inaccuracies on major directories to maintain a clean, authoritative profile.

Encourage and highlight reviews. Platforms like Goodreads, YouTube, and podcast directories allow you to showcase feedback and engagement. Use schema such as Comment or Answer to structure these interactions for machines.

For events, update status and attendance information with schema attributes like eventStatus and eventAttendanceMode. Transparency and accuracy build trust with both users and algorithms.

External authority isn't accidental. It's the result of deliberate outreach, structured data, and ongoing reputation management. The more your work is referenced and respected, the more likely it is to be included in AI-driven answers.

BECOMING A TRUSTED ENTITY IN AI ECOSYSTEMS

AI systems don't just index pages—they recognize entities. Being a trusted entity means your name, brand, or product is mapped in the knowledge graph and referenced by LLMs.

Define your identity clearly. Use schema.org types like Person or Organization, and link to verified profiles with sameAs. Consistency in naming, bios, and images across platforms eliminates ambiguity.

Disambiguate when needed. If you share a name with others, add niche descriptors ("Jordan Lee, healthcare data scientist") and link to relevant topics using schema properties like knowsAbout or hasOccupation. Claim your Google Knowledge Panel and seed data for one if it doesn't exist.

Publish in structured, trusted environments. Books with ISBNs, contributions to Wikidata, and technical content on GitHub or arXiv all help establish you as a distinct entity. These platforms are favored by AI for their reliability and structure.

Entity-first retrieval is the future. LLMs increasingly map answers to people, companies, and products they recognize and trust. If you're not a clearly defined entity, your visibility is limited. If you are, you become a preferred source.

Distribution, structure, and recognition form the triangle of modern authority. When AI systems can find you, understand you, and trust you, your content isn't just indexed—it's relied upon.

KEY TAKEAWAYS

- Discoverability begins with open, structured publishing. Use platforms like Internet Archive, GitHub, and SSRN to ensure your content is accessible to LLMs.
- Schema and consistent metadata make your entity machine-recognizable. Define your authors, organizations, and products clearly across all channels.

CHAPTER 11

Summary and Action Steps

REVIEWING YOUR AEO CONTENT STRATEGY

You've now worked through the core of Strategic AEO Content Development. This section brought you from theory to practice—showing not just what answer engines and AI systems value, but how to build content that gets chosen, cited, and trusted.

You started by learning how pillar content anchors your authority. These aren't just long articles—they're comprehensive, well-structured resources that answer broad questions and support your topic clusters. You learned to create pages that are scannable, semantically clear, and organized for both readers and machines. The goal: become the definitive answer, not just another result.

We focused on answer-first writing. You saw how to lead with the answer, write in a format that's easy to quote, and avoid filler or vague language. Your content should be clear, direct, and ready for AI systems to cite without confusion. Headings, title tags, and metadata all play a role in making your work machine-readable and unambiguous.

Technical optimization was another key theme. Schema markup, JSON-LD, and microdata help clarify your content for AI. We discussed how to implement FAQ, Article, Person, and HowTo schemas, and how to make sure your site is crawlable, mobile-optimized, and free of technical errors. If your content can't be crawled, it can't be cited.

Supporting content and internal ecosystems came next. You learned to map out subtopics, create cluster content that links back to your pillars, and use keyword research tools to uncover conversational queries and "People Also Ask" questions. Internal linking reinforces your authority and helps both users and bots navigate your expertise.

We also covered designing for specific answer contexts. You saw how to structure content for featured snippets, PAA, and voice search—using concise, answer-first sections, natural phrasing, and schema markup to maximize your odds of being surfaced.

Finally, we explored knowledge distribution and external authority. Publishing open-access PDFs, whitepapers, or books (with ISBNs) increases your chances of being included in LLM training sets. Strategic citations, media features, and entity optimization help you build a durable, machine-recognizable presence.

AEO Content Development — Checklist

Pillar Content Creation

- ❐ I have identified the core topics where I want to be the definitive answer.
- ❐ I have created or outlined long-form pillar content that is comprehensive, structured, and semantically clear.
- ❐ My pillar content includes a table of contents, internal anchors, summaries, and clearly labeled sections that can serve as direct answers.

Answer-First Writing Style

- ❐ I write in a clear, concise, answer-first format that prioritizes helpfulness over fluff or marketing jargon.
- ❐ My content avoids vague language, excessive hedging, or filler—it is designed to be quoted by AI, not just skimmed by humans.
- ❐ I have reviewed headings, paragraphs, and title tags to ensure they are unambiguous and machine-readable.

Schema and Technical Optimization

- ❐ I have implemented relevant schema types (e.g., FAQ, Article, Person, HowTo, Book) using JSON-LD or microdata.
- ❐ I have added schema markup to supporting elements such as videos, FAQs, and reviews where appropriate.
- ❐ My site structure is crawlable, mobile-optimized, and technically sound, with correct use of noindex tags, canonical URLs, alt-text, and robots.txt.

Supporting Content and Internal Ecosystems

- ❐ I have mapped related subtopics and created cluster content that links back to each pillar page.
- ❐ My internal linking strategy reinforces topic authority and helps users and bots navigate logically.
- ❐ I use keyword research tools (e.g., SEMrush, Ahrefs, Google Autocomplete) to uncover and target conversational queries and People Also Ask (PAA) style questions.

Designing for Specific Answer Contexts

- ❐ I have structured content to target featured snippets, using definitions, lists, tables, and how-to steps.
- ❐ I embed real user questions as headings and provide concise, direct answers informed by expanded PAA research.
- ❐ My voice-search-optimized content uses simple, conversational phrasing and includes location-specific or action-oriented intent where relevant.

Knowledge Distribution and External Authority

- ❐ I have published or repackaged content as open-access PDFs, whitepapers, or books (with ISBNs where possible).
- ❐ I have made strategic efforts to appear in authoritative spaces such as media features, citations, Wikidata entries, and public repositories.
- ❐ I have claimed authorship and linked my identity and site to structured profiles (LinkedIn, Crunchbase, etc.) using appropriate schema markup.

Next up, how to make sure all of that work actually gets seen. In Part III, we'll dive into visibility — featured snippets, AI interfaces, citations, and everything that gets your content surfaced in the right place, at the right moment, for the right query.

Part III

Maximizing Visibility in AI and Search

The work you've put into building clear, authoritative, and structured content sets the stage. Now, the challenge is to make sure your expertise gets seen—and chosen—across every modern search and answer platform.

In this section, you'll move from asset creation to asset activation. You'll learn how to secure featured snippets, dominate People Also Ask (PAA) sections, and optimize for zero-click results. We'll cover the mechanics of getting your content retrieved by large language models (LLMs) and generative AI, and show you how to influence which assets make it into training data for long-term authority.

Visibility today isn't just about ranking high in search results. It's about **being the answer**—surfaced in snippets, cited in AI summaries, and spoken by voice assistants. To get there, you need to understand how algorithms select content, how retrieval-augmented generation works, and how to structure your information for instant extraction.

You'll see practical strategies for every major channel: web, voice, video, and beyond. We'll dive into compliance, measurement, and technical details that help your work stand out in a crowded, AI-driven landscape. Each chapter offers actionable steps, examples, and frameworks to help you implement these tactics right away.

The goal is simple: make your content the preferred source for both users and AI systems, wherever answers are delivered. Let's move forward and turn your expertise into lasting, visible authority in the answer economy.

CHAPTER 12

Featured Snippets and Zero-Click Optimization

UNDERSTANDING FEATURED SNIPPETS

Featured snippets are the answer boxes that appear at the top of Google's search results, above the traditional blue links. These snippets give users a quick, direct answer without requiring them to click through to a website. You'll see them as definitions, lists, tables, steps, or even short videos. For brands and publishers, winning a snippet means instant authority and maximum visibility.

Google chooses snippet sources based on clarity, directness, and trust. The content must answer the query precisely and come from a site that demonstrates authority. Structured formatting—using bullet points, tables, or step-by-step instructions—makes it easier for algorithms to extract your answer. Schema markup such as FAQPage or HowTo further signals that your content is snippet-ready.

Position zero is powerful. It places your content above all organic results, increasing brand recognition and trust. However, it can also reduce click-through rates, since users may get their answer without visiting your site. To balance this, craft snippets that offer a clear answer but also invite further exploration. Use partial answers or prompts for deeper reading to encourage engagement.

Structured data is essential. Applying FAQ or HowTo schema helps search engines understand your content's intent and structure. This technical layer makes it easier for AI and search engines to feature your content as a snippet.

Featured snippets are more than a ranking win. They represent a shift toward immediate, scannable answers. Align your content with this demand for clarity and speed, and you'll gain both visibility and trust.

Structuring Content to Win Snippets

Winning snippets isn't luck. It's about structure, clarity, and technical precision. Craft your content with these elements in mind. The reward is visibility and authority.

Place snippet-friendly content where it's easy to find. The first few paragraphs of your article are prime real estate. Search engines prioritize this area for quick answers. Keep your language clear and concise. Avoid filler. Every word should serve the answer.

Use headers to guide both readers and algorithms. Structure your content with <h1> for the title, <h2> for main sections, and <h3> for subsections. This hierarchy signals importance and makes your content easy to scan.

Short paragraphs improve readability. Break ideas into chunks of one to three sentences. Use bold text sparingly to highlight key phrases. This draws attention from users and search engines alike.

Implement structured data for rich snippets. Add FAQ or HowTo schema to your content. For example, wrap questions and answers in FAQ schema. This tells search engines exactly how to display your information.

Tables and lists are especially effective. Present data in a scannable format for comparisons, steps, or itemized answers. AI systems favor this clarity.

Mobile design matters. Ensure your content adapts to smaller screens. Search engines prioritize mobile-friendly pages. A responsive layout keeps your snippet chances high.

Avoiding Over-Optimization

Targeting snippets can boost visibility, **but overdoing it risks losing organic traffic.** Search engines may flag excessive optimization as manipulative, leading to algorithmic adjustments or manual reviews that reduce your content's reach.

Balance is key. Some queries deserve snippet-friendly answers. Others require deeper, authoritative content. Prioritize based on intent. For example, "how to tie a tie" suits a quick snippet. "History of neckwear" demands a comprehensive guide.

Authority matters. Thin or overly optimized content fails E-E-A-T (Experience, Expertise, Authoritativeness, Trustworthiness) standards. Build trust with expertise and depth. Snippets should complement, not replace, substantive information.

Technical choices matter too. Use redirects to preserve link authority during content updates. Tools like Google Search Console help spot over-optimization issues early.

Opportunity scoring guides decisions. Target snippets for high-priority keywords aligned with user needs. Avoid forcing snippets where they don't fit.

The goal is sustainable authority. Optimize thoughtfully, and your content will thrive in both snippets and organic results.

KEY TAKEAWAYS

- Featured snippets on Google create instant authority. Winning position zero places your content directly into AI overviews, voice assistants, and zero-click answers, dramatically increasing brand visibility.
- Structure drives snippet success. Clear headings, concise paragraphs, scannable lists, and properly applied schema markup make your content easier for algorithms to extract and surface.

CHAPTER 13

Dominating People Also Ask (PAA)

How the PAA Algorithm Works

The "People Also Ask" (PAA) section is one of the most dynamic features in search. It expands as users interact, revealing a branching tree of related questions. Each click triggers more questions, mapping user curiosity in real time. If your content answers these questions well, you can surface repeatedly within a single session.

PAA questions follow a hierarchy. The first layer is broad, often matching the main query. Each subsequent click reveals more specific or tangential questions. Google's algorithm selects PAA entries based on clarity, semantic relevance, and authority. Pages with structured, concise answers—especially those marked up with schema—are more likely to be chosen.

The PAA box isn't static. It adapts to user behavior and emerging trends. As more people search and click, new questions appear and old ones fade. Understanding this logic helps you align your content with what users and algorithms want.

Systematic PAA Content Strategy

Use PAA like a content roadmap: identify the questions people actually ask, structure your articles around them, and answer each one so clearly that Google has no choice but to surface you in the box.

Start by researching the questions that appear most often in your niche. Tools like AlsoAsked, Ahrefs, and SEMrush can map question clusters and uncover gaps in your coverage. Manual expansion of the PAA box reveals additional queries and variations.

Embed high-value PAA questions as H2 or H3 headings within your content. Each heading should be followed by a direct, self-contained answer. Aim for 40–50 words per response. Avoid filler and sales language. Focus on being clear and precise.

Semantic twins—variations of the same question—are also helpful. For example, "How does schema markup help SEO?" and "Why use schema markup for search?" target similar intent. Cover both, but reword naturally to avoid repetition.

Create FAQ sections for related questions. Use FAQ schema to mark up these blocks, making your answers machine-readable. However, don't rely solely on end-of-page FAQs. Google prefers answers that appear in context, embedded within the main body of your content.

Internal linking strengthens your PAA ecosystem. Link from supporting articles to pillar pages, and between related questions. This structure signals depth and authority, increasing your chances of being selected for multiple PAA spots.

Monitor performance with analytics and rank-tracking tools. Identify which questions drive the most engagement, then refine your content to target high-value PAA opportunities.

Technical Execution for PAA Wins

Treat PAA like an engineering problem as much as a content one: the right schema, crawlable layout, and mobile-first performance turn good answers into answers search engines can actually see, trust, and promote.

Technical precision is essential. Use JSON-LD to mark up FAQ sections, ensuring each question and answer pair is clearly defined. Validate your structured data with tools like Google's Rich Results Test. Address any errors or mismatches promptly.

Clean site architecture supports discoverability. Avoid blocking essential resources in robots.txt—your set of rules for bots and crawlers that browse your website. Ensure all answers are crawlable and not hidden behind tabs or accordions that search engines can't access. Use jump links for easy navigation, especially on long pages.

Mobile-first design is critical, this means making sure your website displays properly on smartphones and tablets. PAA results are often surfaced on mobile devices. Optimize layouts for small screens and fast load times. Avoid intrusive pop-ups or overlays that disrupt the user experience.

KEY TAKEAWAYS

- PAA offers compounding visibility. Winning one PAA box can lead to multiple appearances across related queries, amplifying your reach.
- Research and structure are essential. Use tools to map question clusters, embed high-value queries as headings, and answer them directly within the content.

CHAPTER 14

Voice Search Optimization—Becoming the Spoken Answer

How Voice Search Works Behind the Scenes

Voice search is now a daily habit for millions. People ask questions on their phones, smart speakers, and even in their cars. They want quick, spoken answers—not a list of links. **When you optimize for voice, you're aiming to be the single answer that gets read aloud.**

Voice assistants like Google Assistant, Siri, and Alexa use a mix of data sources. They pull from featured snippets, structured data marked up with schema, Google Business Profiles, and local search results. For many queries, especially those about facts or how-tos, assistants rely on content that is clear, well-structured, and easy to speak.

These systems use retrieval and ranking logic. First, they find a set of possible answers. Then, they choose the one that is most accurate, current, and easy to deliver. If your content is the best fit, it gets spoken. If not, it's skipped. There are no second chances in voice search.

Mobile readiness, fast load times, and semantic clarity all play a role. Voice search engines want to be confident your answer is correct and easy to understand. That means technical optimization and content clarity work together.

Writing for Voice: Tone, Clarity, and Word Choice

Writing for voice is different from writing for the screen. Spoken answers need to be simple, direct, and conversational. If your content sounds awkward when read aloud, it won't get chosen.

Use plain language. Avoid jargon and long, complex sentences. Keep sentences short—often four to ten words. Read your answers out loud as you edit. If you trip over a phrase, rewrite it.

Contractions help. Say "you'll" instead of "you will." This makes the tone feel natural, like a real conversation. Aim for a neutral, helpful voice. Avoid sales language or dramatic flair. Get to the point quickly, and let the answer stand on its own.

Most voice answers are 40 to 50 words. That's your window. Lead with the answer, then add a supporting detail if needed. Don't bury the answer in an introduction. If a user asks, "How do I reset my router?"then the first sentence should be the step-by-step process.

Punctuation and grammar matter, but rhythm and clarity matter more. If a sentence is too long to say in one breath—split it. If a word sounds strange or is hard to pronounce, replace it.

The One-Answer Rule

Voice search is winner-take-all. There's no second result, no scrolling, and no "see more." Either your content is chosen and spoken, or it isn't heard at all.

This is the **One-Answer Rule**. To win, your answer must be:

- **Factually accurate** and up to date.
- **Clear** and easy to read aloud.
- **Confident**—free from hedging or filler.

Even small issues can push your content out of contention. A confusing phrase, a dated fact, or a mismatch in tone can cost you the answer spot. Voice search engines update their answers frequently, so content freshness is key.

When you write for voice, ask yourself: "Is this the answer I would want a machine to say out loud if it only said one?" If it isn't, revise until it is.

Tuning for Local and Transactional Voice Queries

Most voice queries are about real-world needs—finding a business, ordering food, or getting directions. Local and transactional queries are high-value and time-sensitive. If your business isn't optimized for these, you're missing out.

Start with your Google Business Profile. Make sure your hours, phone number, and address are accurate and consistent. Voice assistants often pull from this data for local queries.

Use LocalBusiness schema on your site. Include details like business hours, service area, and contact information. If you offer services, add schema for makesOffer or hasService. This helps machines understand what you do and where you are.

Phrase your content for action. People say things like "call now," "order online," or "book today." Make these actions easy to find and use. Include clickable phone numbers and clear calls to action.

Customer reviews matter. Positive reviews, especially those with relevant keywords, help voice engines trust your listing. Encourage customers to leave feedback and respond to reviews promptly.

Voice queries are often urgent. Make sure your site loads quickly, works on mobile, and gets users to the answer or action fast.

Testing and Diagnosing Voice Visibility

Voice search performance isn't always obvious. There's no dashboard that says, "You're the answer." But you can test and diagnose your visibility.

Start by asking real questions on voice devices. Use Google Assistant, Siri, or Alexa to see what answer gets spoken. If your content is read aloud, you're doing well. If not, check for featured snippets on desktop or mobile. Most voice answers come from these.

Review your schema markup. Use Google's Rich Results Test or schema validators to catch errors. Missing or broken schema can prevent your content from being selected.

Try voice simulation tools like Jetson.ai or Google Assistant Simulator. These can show how your answers might be interpreted in spoken contexts.

Watch your analytics for indirect signals. If you see more mobile or zero-click traffic, it may mean your content is being used in voice search. Traffic from long-tail queries with no clicks can also be a clue.

Voice search requires ongoing testing. Regularly check your answers and update them as needed.

VOICE PERSONAS AND USE-CASES

Different people use voice search in different ways. Understanding these personas helps you tailor your content.

- **The Commuter**: Needs quick, location-based answers while driving. Short, clear responses work best.
- **The Smart Speaker User**: Asks for facts, recipes, or reminders at home. Natural language and clear lists help.
- **The Multitasker**: Wants the fastest route to an answer while doing other things. Brevity is key.
- **The Local Searcher**: Looks for nearby services or businesses. Hyperlocal answers and up-to-date contact info are essential.

Write your answers with these use-cases in mind. Match your phrasing, length, and structure to how and where people are asking.

Key Takeaways

- Voice search is winner-take-all. Only one answer gets spoken, so clarity and accuracy are critical.
- Write in plain, conversational language. Short sentences and contractions sound natural when read aloud.

CHAPTER 15

Getting Retrieved by LLMs and Generative AI

Understanding Retrieval-Augmented Generation (RAG)

Retrieval-augmented generation (RAG) is changing how large language models (LLMs) deliver answers. Instead of relying only on what was learned during training, RAG lets AI models pull in up-to-date information from external sources in real time.

When a user asks a question, the model searches trusted sites, APIs, or its own indexed knowledge, then combines that data with its generative capabilities to create a relevant, current answer.

Often-times, the preferred web-search API is Google, meaning that AI is just getting Google search results for RAG. This means that your first priority for getting retrieved by LLM or AI-based searches is still SEO.

For your content to be chosen in this process, it must be both discoverable and machine-readable. Consistency matters. LLMs favor data that aligns across platforms—names, facts, and structured signals must match everywhere they appear. Structured data, like schema markup, acts as a roadmap for retrieval, clarifying what your content represents and how it should be used.

RAG rewards clarity and authority. If your content is ambiguous, outdated, or inconsistent, it's likely to be skipped. To qualify as a retrieval candidate, you need to focus on precision, semantic structure, and up-to-date information.

Becoming a "Retrieval Candidate" for RAG

Visibility in retrieval systems starts with organic search (SEO). Content that ranks on the first or second page of search results is more likely to be indexed and retrieved by LLMs.

But ranking isn't enough. Your content structure, domain's reputation, the quality of your backlinks, and the authority of your content all influence whether you're surfaced in real-time answers.

LLMs and RAG systems don't just look at whole pages—they index and retrieve passages. Structure your content so each section can stand alone as a complete answer: use descriptive H2/H3 headings, start with a direct one-sentence response, then add supporting detail, bullets, or tables. This "chunk-friendly" structure makes it easier for retrieval systems to lift and cite your work accurately.

Page-level trust is critical. Make sure your content is free from technical errors, such as broken links or missing schema. Use structured data to clarify the purpose and context of each page. This helps AI systems extract and cite your work accurately.

Align your writing with how people phrase their questions. LLMs understand semantics, not just keywords. Use natural, conversational language that mirrors the way users speak and search. This increases the odds your content matches the intent behind queries.

Monitor how LLMs cite your work. If you notice your content isn't being referenced, adjust your phrasing, structure, or schema. Treat this as a feedback loop—refinement improves your retrieval potential.

Technical optimization supports retrieval. Maintain a crawlable, mobile-friendly site. Use 301 redirects to preserve link authority during updates. Clean, accessible pages are easier for AI to process and retrieve.

Structuring Content for Real-Time Retrieval

To maximize your chances of being retrieved, structure your content for easy extraction. Use single-topic paragraphs that can stand alone. Each should include enough context to make sense out of order, since LLMs often cite fragments or summaries.

Attribution helps. Phrases like "According to [notable author name]" or "Research shows" signal authority and make your content ready for citation. Avoid excessive first-person language, which can feel less objective to both users and algorithms.

Neutrality is important. Focus on facts and expertise, not brand-centric language. Let your content speak for itself with clear, evidence-based statements.

Leverage structured data. Schema markup clarifies relationships between ideas and guides AI to the most relevant sections. Mark up entities, authors, and key facts to make them machine-readable.

Avoid duplication. Unique, focused paragraphs outperform recycled or overly generic text. Each section should reinforce your authority and improve retrieval odds.

Formatting is about more than aesthetics. It's about making your content machine-friendly. When you structure your work for clarity and extraction, you become a go-to source for answers.

Key Takeaways

- RAG enables LLMs to pull real-time data from trusted sources, making your content eligible for dynamic inclusion in AI-generated answers.
- Consistency across platforms—names, metadata, and structured signals—reduces ambiguity and increases retrieval confidence.

CHAPTER 16

Influencing LLM Training Data Inclusion

WHY TRAINING DATA IS A STRATEGIC ASSET

Training data sits at the core of every large language model (LLM). When your content is included in a model's training set, it shapes how that AI understands your field for years to come.

Unlike retrieval, which is dynamic and fleeting, training data inclusion embeds your expertise directly into the model's foundation. **This is how your knowledge becomes part of the answers users receive—sometimes long after your content is published.**

Think of training data as the curriculum for AI. The more often your work appears in trusted, structured, and accessible sources, the more likely it is to be learned and echoed by future models. This helps you build a long-lasting presence.

Models train in cycles. Once your content is included, it can inform responses for months or even years. That kind of influence compounds over time, making training data inclusion the deepest form of authority in the answer economy.

HOW CONTENT GETS TRAINED INTO MODELS

LLMs are trained on vast, curated datasets pulled from across the public web. Not every page makes the cut. Models like GPT, Claude, and Gemini rely on sources that are open-access, well-structured, and widely cited.

- **Common Crawl** is a major feeder, scraping billions of pages but filtering for trust, clarity, and accessibility.
- **ISBN registration** turns your writing into a recognized artifact, increasing its weight and discoverability.

- **Books, academic papers, and technical documentation** are especially valuable. Content published through the Internet Archive, Google Books, Open Library, arXiv, or SSRN is more likely to be included.
- **Technical documentation** on platforms like GitHub or Read the Docs is also favored. Markdown files, README guides, and API references are easy for machines to parse and ingest.

The more structured and accessible your content, the higher its odds of training inclusion.

Accessibility is critical. Content behind paywalls, login screens, or heavy JavaScript is often skipped. Models need to crawl, process, and legally use your work. Open licensing, semantic formatting, and clear metadata all improve your chances.

Citation networks matter. Content that's referenced by other trusted sites, academic papers, or industry blogs is more likely to be selected for training. The more your work is cited, the more it stands out during dataset curation.

STRATEGIC PUBLISHING AND TRUST SIGNALS

Where and how you publish makes a difference. LLMs favor open-access repositories and platforms designed for stability and structure. The Internet Archive, arXiv, SSRN, and GitHub are prime examples.

Format is as important as location. HTML and Markdown are ideal. PDFs are acceptable if they're tagged semantically and not just scanned images. Use schema.org markup or similar standards to add semantic cues. The less clutter, the better.

Metadata anchors your content. Consistent author names, clear titles, publication dates, and headings help models organize and interpret your work. Avoid ambiguity in your bylines or branding. Unified signals across your publishing history build a strong, machine-recognizable identity.

Domain authority boosts inclusion odds. Government, educational, and long-established domains are prioritized. If you don't control one, aim to be cited by one. The citation patterns that matter to researchers also matter to AI training pipelines.

Crawlability is non-negotiable. Keep your robots.txt open for public content. Maintain a current sitemap. If a model can't reach your content directly, it won't include it—no matter how valuable the information. This includes what visibly displays first, so avoid pop-ups, excessive JavaScript, or interstitials that block text.

Language and tone matter, too. Write with clarity and authority. Avoid excessive self-promotion or filler. LLMs are trained to look for direct, well-supported claims.

Strategic publishing is a trust loop. Good formatting leads to better indexing, which leads to more citations, which increases domain authority, and in turn, raises your chance of being included in the next training cycle.

THE CITATION MAGNET METHOD

Citations are gravity for training data. When other sites, papers, or books reference your work, it signals to curators and algorithms that your content is trusted and influential. This not only improves your retrieval odds but also your training inclusion.

To attract citations:

- **Publish** original research, guides, or frameworks that others want to reference.
- Use clear, citable statements and **attribute** data or claims.
- **Distribute** your work across platforms where academics, journalists, and industry experts look for sources.
- **Engage** with communities in your field by contributing to open-source projects, academic discussions, or public datasets.

Track your citations using tools like Google Scholar, Semantic Scholar, or backlink analysis platforms. The more your work is cited, the more it stands out for both retrieval and training.

KEY TAKEAWAYS

- Training data inclusion is the deepest form of authority in AI. It shapes how LLMs answer questions for years, not just days.
- Models prioritize open, structured, and accessible content. Publish on trusted platforms, use clean formats, and keep metadata consistent.

CHAPTER 17

Local Answer Engine Optimization—Winning the "Near Me" Query

Why Local AEO Is Different

If you are doing answer engine optimization for a business or organization with a physical location, or a specific area of operation, local AEO is essential.

Local SEO and AEO is important because modern searches often include location data from the user, typically IP Geolocation or GPS from a phone. **This means that most searches that have a local intent will automatically have the location added to the search.**

Searches with local intent include keywords that indicate people want answers that are close, open, and ready to help. Whether someone asks, "best pizza near me," "urgent care open now," or "hardware store in [city]," answer engines and AI platforms now deliver results that blend relevance, proximity, and trust.

Winning these queries is about more than showing up on a map. You need to be the most locally relevant answer—spoken by voice assistants, surfaced in mobile search, trusted by AI-driven interfaces.

Optimize Your Google Business Profile Like Core Content

Your Google Business Profile (GBP) is often your first—and sometimes only—impression in local results. Treat it like a pillar page. Every detail matters.

Start with accuracy. Ensure your business name, address, and phone number (NAP) are consistent everywhere. Even a small mismatch can confuse algorithms and erode trust. Double-check your hours, including holidays and special events. Outdated info leads to lost opportunities.

Write your business description with intent. Use natural language that matches how people ask about your services. "Family dentist open late in Austin" is more effective than a generic tagline. Include service areas, specialties, and unique selling points.

Visuals build trust. Upload recent, high-quality photos of your storefront, team, and products. Update regularly. Active profiles signal relevance to both users and search engines.

If you offer appointments or bookings, link directly to your scheduling tool. Google often pulls this into voice and mobile search, making it easier for users to take action.

Review your business categories. Choose those that match how locals search, not just what fits in a directory. For example, "back pain treatment" might connect better than "chiropractor."

Keep your GBP fresh. Post updates, events, or offers. Activity signals engagement and can push your listing higher in local results.

Build Content Around Localized Questions

Local AEO goes beyond adding a city name to your title. It means answering the real questions your community asks.

Start by researching. Use Google's "People Also Ask," autocomplete, and tools like AlsoAsked or AnswerThePublic to uncover natural-language, location-specific queries. Analyze your own analytics for long-tail phrases that bring in local traffic.

Create content that directly addresses these questions. Service area pages, city-specific FAQs, and blog posts about local events or seasonal needs all help. For example, "Best HVAC maintenance tips for Phoenix summers" targets a real concern and can win both featured snippets and voice answers.

Mention local landmarks, neighborhoods, or events. Show you're part of the community. This builds trust with users and signals hyperlocal relevance to algorithms.

Link local pages back to your pillar content. Internal linking strengthens your authority and gives search engines more context about your expertise in specific locations.

Manage Citations and Listings Consistency

Citations are the backbone of local trust. Search engines scan the web for mentions of your NAP data. Inconsistencies weaken your authority and can prevent your business from appearing in "near me" results.

Claim and verify your business on major directories: Yelp, Bing Places, Apple Maps, and industry-specific platforms. Use tools like Moz Local, BrightLocal, or Yext to audit and fix discrepancies.

Ensure every listing has the same business name, address, and phone number as your website and GBP. Check business categories, links, and hours of operation for accuracy. Outdated or mismatched data sends mixed signals and can hurt your rankings.

Review your listings regularly. Mergers, moves, or rebrands can leave old data floating online. Consistent, up-to-date citations reinforce your legitimacy across every channel.

Local Video Signals: Walk-and-Talks in Real Places

Short "walk-and-talk" clips – filmed on a phone in real locations – create strong local signals for both humans and algorithms. When the background shows recognizable landmarks, storefronts, or neighborhoods, platforms like TikTok, Instagram, and YouTube can infer geography visually, not just from captions or tags.

For local AEO, treat these clips as another structured signal:

- **Film in front of recognizable local spots** - main streets, landmarks, partner businesses, community centers.
- **Mention the city and neighborhood** out loud as you talk ("Here in Oxnard, in the Marina district...").
- **Keep them 30–60 seconds** and focus on one problem, one tip, or one story.
- **Whenever possible, feature local clients** and let *them* speak on camera about a specific outcome.

After publishing:

- **Embed the best clips** on your local service pages and location pages.
- **Add VideoObject schema** that includes your city and service area in the description.

The combination of LocalBusiness schema, consistent profiles, and locally grounded video gives answer engines multiple layers of evidence that you are the relevant expert for that geography.

TURN CUSTOMER REVIEWS INTO LOCAL AEO SIGNALS

Reviews are more than reputation—they're ranking factors. Search engines and AI systems look for recent, specific, and relevant feedback to decide which businesses to surface.

Encourage customers to leave reviews after a positive experience. Make it easy by sharing direct links or QR codes on receipts and follow-up emails.

Ask customers to mention what they had done and where. "Quick oil change at your downtown Seattle shop" is more valuable than "Great service."

Respond to every review, positive or negative. Engagement shows you're active and attentive. Google and other platforms reward businesses that interact with their customers.

Avoid fake or incentivized reviews. Authenticity matters. Manipulative tactics are easy to spot and can lead to penalties or removal from local listings.

Mobile and On-the-Go Experience

Most local searches happen on mobile devices. Your site needs to load fast on smartphones and make key actions obvious.

Test your site with Google's PageSpeed Insights. Compress images, reduce scripts, and prioritize mobile usability. Place your phone number and address at the top of the page. Use large, tappable buttons for calls, directions, or bookings.

Design for real-world urgency. A mobile visitor wants to know if you're open, how to get there, and how quickly you can help. Don't bury this information behind menus or long paragraphs.

Implement LocalBusiness and Action schema to help search engines interpret your site's intent. Structured data supports both mobile and voice accessibility.

Test your site on different devices and network speeds. What works on a new iPhone may not on an older Android. Simulate real-world conditions to ensure a smooth experience for everyone.

Key Takeaways

- Local AEO is about being the answer for real-world, location-based queries—especially in voice and mobile search.
- Treat your Google Business Profile as a living pillar asset. Keep NAP data, hours, and categories current and consistent.

CHAPTER 18

Optimizing for Search in AI Interfaces

UNDERSTANDING SEARCH IN AI INTERFACES

AI interfaces, from chatbots like ChatGPT to generative search platforms and voice assistants, are transforming how users discover information. These systems don't just list results—they synthesize, summarize, and deliver answers in real time. The mechanics behind these platforms are different from traditional search engines. **Queries are interpreted conversationally, sources are chosen for clarity and trust, and the answer is often presented as a single, authoritative statement.**

You'll notice that **AI interfaces often blend retrieval with generation**. They pull from live web data, pre-trained knowledge, and structured sources. Each platform has its own logic for parsing queries and selecting what to cite. Understanding these nuances is key to making your content visible and preferred in these environments.

HOW SEARCH WORKS IN CHAT AND VOICE INTERFACES

Monitoring your performance in AI interfaces is still evolving. Manual prompt testing, analytics, and brand monitoring tools can help you spot trends and refine your strategy.

Chat-based search is dynamic. Users type or speak questions in natural language, often expecting a direct, context-aware answer. AI agents analyze the query, fetch relevant data from trusted sites, and assemble a response. Retrieval-augmented generation (RAG) is common—AI fetches live information, then summarizes it using its language model.

Testing prompts is essential. Try asking the same question in different ways. You'll see how phrasing changes the answer. Track which sources are cited. Note if your content appears, and how it's paraphrased or quoted.

Voice search adds urgency. Users want a single, spoken answer, not a list. AI pulls from structured data, featured snippets, and business profiles. If your content is clear, current, and easy to read aloud, it's more likely to be chosen.

Geo-specific queries matter. AI adapts results based on the user's location. Local content, optimized for "near me" and city-specific searches, is critical for appearing in these contexts.

Structuring Content for AI Search Selection

For AI search selection, clarity is the top priority. Write in plain, direct language. Avoid jargon unless your audience expects it. Each paragraph should stand alone, making it easy for AI to extract and cite.

Use schema markup to clarify relationships and context. FAQ, HowTo, LocalBusiness, and Article schemas help AI systems parse and prioritize your content. Structured data signals intent and authority, increasing your odds of being selected.

Frame your headlines and subheadings for intent. Match the way real users phrase their questions. Aligning your titles with common prompts helps AI recognize your content as a relevant answer.

Cite reputable sources. Link to high-quality references and demonstrate expertise. This builds trust with both users and algorithms.

Keep your content fresh. Update older posts with new insights and data. Recency is a ranking factor for AI-driven platforms.

Remove barriers to access. Avoid paywalls, excessive ads, or code bloat that could block crawling. Ensure your site is mobile-friendly and loads quickly.

Cross-Platform Visibility Tactics

Visibility in AI interfaces isn't limited to one channel. Your content needs to be optimized for Google SGE, Bing Copilot, Perplexity, voice assistants, and more. Each platform may have unique requirements, but some tactics apply everywhere:

- **Apply schema markup**: Use FAQ, HowTo, Book, VideoObject, and LocalBusiness schemas where appropriate.
- **Submit sitemaps and URLs**: Keep your XML sitemaps current. Make sure to cover both Google and IndexNow.
- **Monitor performance**: Use Google Search Console, Bing Webmaster Tools, and prompt testing to track where and how your content appears.
- **Optimize for speed and accessibility**: Fast, clean, and crawlable pages are more likely to be indexed and cited.
- **Claim and update business profiles**: Ensure your Google Business Profile and other directory listings are accurate and current.
- **Encourage reviews and engagement**: User feedback and interaction signals boost trust and increase your odds of being cited in AI-driven results.

Tailor your approach for each major platform. For YouTube, focus on keyword-rich titles, transcripts, and VideoObject schema. For books, use Book schema and maintain consistent ISBN metadata. For local businesses, prioritize LocalBusiness schema and review management.

Key Takeaways

- AI interfaces blend search and conversation. They select answers based on clarity, authority, and structured data.
- Test your content with real prompts. Manual and automated checks reveal how your work is cited and summarized.

CHAPTER 19

Channel-Specific Strategies

WHY CHANNEL-SPECIFIC OPTIMIZATION MATTERS

Every platform—whether it's local search, YouTube, Amazon, or a book database—has its own rules, technical requirements, and user expectations. Answer Engine Optimization (AEO) is not one-size-fits-all. **To maximize your visibility and authority, you need to tailor your approach for each channel where your audience searches, asks, or consumes answers.**

You'll notice that what works for Google's main search results might not work for YouTube's video carousel, a voice assistant's spoken answer, or a book's knowledge panel. Channel-specific strategies ensure your content is not just present, but preferred—surfaced, cited, and trusted by both users and AI systems in every relevant context.

LOCAL SEO FOR CONVERSATIONAL AND VOICE SEARCH

For local businesses, AI and voice search are the new "front door"—if assistants and agents can't understand who you are, where you are, and what you do, they'll send customers to someone else.

Local visibility starts with structured data. Use LocalBusiness schema to clarify your business name, address, phone number, and service area. Consistency is critical. Make sure your NAP data matches across your website, Google Business Profile, directories, and social platforms.

Update your Google Business Profile regularly. Keep hours, services, and photos current. Voice assistants and AI agents often pull directly from this profile to answer "near me" queries. The more accurate and complete your profile, the more likely you are to be chosen as the answer.

Target conversational "near me" searches. People ask for businesses the way they speak: "best sushi near me," "urgent care open now," or "coffee shop with Wi-Fi." Build content that answers these queries directly. Create FAQ pages, service area content, and blog posts that address local needs in natural language.

Encourage and manage reviews. Positive, detailed feedback with relevant keywords helps both search engines and AI systems understand your strengths. Respond to every review—engagement signals trust and activity.

List your business on reputable directories. Audit for consistency and accuracy. Inconsistent citations confuse algorithms and can undermine your local authority.

Monitor your presence in AI and voice search. Test how your business appears in voice results, and adjust your strategy based on what you find. Local SEO for AI is about clarity, consistency, and trust.

Video SEO and YouTube Optimization

YouTube is a primary source for video answers in Google and AI-powered search.

To optimize for Video SEO and AEO:

- **Use keyword-rich, descriptive titles.** Place the most important words at the beginning.
- **Add timestamps in your video descriptions.** Segmenting content helps users and search engines find what matters most.
- **Include accurate transcripts or captions.** These make your content accessible and give search engines more text to crawl.
- **Link to pillar content in your video descriptions.** Drive viewers to your website and reinforce your topical authority.
- **Embed videos on relevant pages.** This increases engagement and signals quality to search algorithms.

- **Use VideoObject schema to mark up embedded videos.** This improves how they appear in search and AI summaries.

Remember to analyze competitors. Study their top-performing videos for keyword use, engagement tactics, and structure. Adapt insights to your own strategy.

YouTube optimization is not optional. It's a core part of modern AEO and multiplies your chances of being cited or surfaced in generative search.

ISBN, Book Metadata, and Knowledge Panels

Books and long-form content are durable authority assets. Registering an ISBN turns your work into a recognized entity. Search engines and AI systems use ISBNs and structured metadata to display books in knowledge panels and answer boxes.

- **Register** your book through official agencies like Bowker or Nielsen.
- Complete all **metadata**: title, author, publisher, description, and publication date. Consistency across Amazon, Google Books, and Goodreads is essential.
- Use Book **schema** on your website. Specify ISBN, author, and publisher to help search engines display your book in rich results.
- Submit your book metadata to **Google Books** through their Partner Program.
- Claim your **knowledge panel** as an author or publisher. Update cover art and descriptions for accuracy.

Monitor related ISBNs in your field. Analyze their metadata and schema to find gaps or opportunities for your content.

Books are more than content—they're searchable, long-term authority signals. Treat them with the same SEO care as your website.

TECHNICAL PRECISION FOR CHANNEL SUCCESS

Technical details unlock visibility everywhere. For all channels:

- **Use structured data** (schema) tailored to the content type: LocalBusiness, VideoObject, Book, FAQ, HowTo, etc.
- **Ensure your pages are crawlable.** Avoid paywalls, excessive ads, or code that blocks indexing.
- **Submit sitemaps and use IndexNow** for instant updates on Bing and Yandex.
- **Keep canonical tags accurate** to consolidate ranking signals and avoid duplicate content.
- **Test your content on multiple devices and platforms.** What works for desktop may fail on mobile or voice.

Cross-platform visibility is not luck. It's the result of structured data, technical hygiene, and ongoing monitoring.

KEY TAKEAWAYS

- Every channel—local, video, book, or voice—demands a tailored optimization strategy. Generic approaches miss opportunities for authority and discovery.
- Local SEO hinges on clarity, NAP consistency, and conversational content. Google Business Profile is your local pillar asset.

CHAPTER 20

Part III Summary and AEO Visibility Checklist

MAXIMIZING VISIBILITY—WHAT YOU'VE ACHIEVED

Part III guided you through making your answers visible and unmissable across every channel where users and AI systems seek information.

This section was about more than ranking. It was about ensuring your expertise gets surfaced, cited, and trusted—whether in search results, answer boxes, voice assistants, or AI-generated responses.

You learned how to engineer your content for instant extraction. We broke down featured snippets—what Google and AI models look for, how to structure your answers for "position zero," and how to balance clarity with depth so you win visibility without sacrificing engagement.

You then explored People Also Ask (PAA), learning to uncover real user questions, map out intent clusters, and write concise, semantically clear answers that machines can lift directly. Your content now meets users at every follow-up, not just the initial query.

We moved beyond the SERP, showing you how LLMs and generative AI platforms retrieve and cite content. You learned to optimize for real-time extraction, structure for retrieval-augmented generation, and format your work to be cited as a trusted authority in ChatGPT, Bing Copilot, Perplexity, and more.

You also discovered how to influence LLM training data. Some content is retrieved at runtime, but the most influential assets are learned and embedded during model training. You now know how to increase the odds of your books, PDFs, and open-access posts being included in these foundational datasets.

Because AI-driven discovery is multi-channel, **you expanded your toolkit for cross-platform visibility.** You learned to optimize for Google, Bing, voice assistants, YouTube, Google Books, and local search. You now know how to structure video transcripts, manage book metadata, and claim ISBN-level visibility.

Part III was about answer distribution—ensuring your expertise appears wherever users search, speak, or let AI do the asking. With these strategies, your content becomes unmissable—surfaced, cited, and trusted across the evolving landscape of search and AI.

AEO VISIBILITY — CHECKLIST

Featured Snippets and Zero-Click Results

- ❒ Structure content to answer directly after headings, using concise paragraphs, lists, or tables.
- ❒ Use FAQ, HowTo, and QAPage schema to reinforce snippet eligibility.
- ❒ Balance snippet clarity with deeper content that encourages further engagement.

People Also Ask (PAA) Optimization

- ❒ Research and embed high-value, real user questions as H2 or H3 headings.
- ❒ Provide direct, self-contained answers in 40–60 word blocks.
- ❒ Cover semantic twins—variations of core questions—without duplicating content.

Voice Search Optimization

- ❒ Write concise, conversational answers that sound natural when spoken aloud.
- ❒ Lead with the key fact or instruction within 40–50 words.
- ❒ Use schema and accurate business data to support voice and local results.

LLM and Generative AI Retrieval

- ❒ Ensure your content ranks in the first two pages of organic results for target queries.
- ❒ Use structured data to clarify entities, context, and relationships.
- ❒ Write in natural, conversational language that matches user intent.

Influencing LLM Training Data

- ❐ Publish open-access, well-structured assets (PDFs, books, blog posts) on trusted repositories.
- ❐ Register ISBNs and maintain consistent metadata for books and long-form content.
- ❐ Contribute to platforms like Internet Archive, arXiv, SSRN, and GitHub for technical topics.

Channel-Specific Optimization

- ❐ Optimize Google Business Profile and local listings for "near me" and voice queries.
- ❐ Use VideoObject schema, transcripts, and keyword-rich titles for YouTube and embedded video.
- ❐ Tailor metadata and schema for books, ensuring ISBN and author consistency.

Technical and Measurement Best Practices

- ❐ Keep your site crawlable, mobile-optimized, and free of technical errors.
- ❐ Submit updated sitemaps and use IndexNow for instant indexing where supported.
- ❐ Track featured snippet wins, PAA placements, and AI citations using analytics and monitoring tools.

Next up, in Part IV, we'll shift from execution to **optimization and longevity**: how to monitor your AEO performance, debug technical issues, keep your content updated, and ensure your visibility compounds over time — even as AI search evolves.

PART IV:

Optimization, Measurement, and the Future

You've learned how to build authoritative content and make it visible across search engines, answer boxes, and AI-driven interfaces. Now, Part IV shifts the focus to long-term growth, adaptability, and maintaining your edge as the answer economy evolves.

This section is about sustaining and scaling your AEO strategy. You'll see how to measure what matters, stay compliant with changing policies, and build processes for continuous improvement. We'll cover how to audit your assets, refine your technical foundations, and monitor your presence across both traditional and AI-powered channels.

You'll also explore advanced tactics for reinforcing your authority—expanding your entity footprint, deepening your knowledge network, and adapting to new retrieval and training trends as AI systems change. Each chapter provides actionable steps to help you future-proof your content, maintain trust, and ensure your expertise remains visible, cited, and chosen—no matter how search and AI platforms evolve.

Part IV is your playbook for resilience and growth. The answer economy never stands still. With these strategies, you'll be ready to adapt, improve, and lead—today and tomorrow.

<u>CHAPTER 21</u>

Monitoring AEO Performance Across Platforms

WHY MONITORING MATTERS IN MODERN AEO

Visibility is never guaranteed. Search engines, answer engines, and AI platforms update constantly. What wins today may fade tomorrow. Monitoring your Answer Engine Optimization (AEO) performance is not just a technical task—it's a strategic necessity. **You need to know where your answers are surfacing, how they're being cited, and where you're losing ground. Without this insight, you risk falling behind as algorithms and user behaviors evolve.**

AEO monitoring goes beyond tracking rankings. You're looking for citations in featured snippets, appearances in People Also Ask (PAA), voice search results, and references in AI-generated responses. You'll also want to watch for indirect signals: zero-click traffic, branded queries, and mentions in summaries or chat interfaces. The goal is to see the full picture—across every channel where your expertise could be surfaced.

QUANTIFYING GOOD USER EXPERIENCE

Good user experience, or how good of an experience people will have using your website, is measured by page experience signals.

The three main page experience metrics set by Google—Largest Contentful Paint (LCP), Interaction To Next Paint (INP), and Cumulative Layout Shift (CLS)—impact rankings and visibility for both SEO and AEO.

- **Largest Contentful Paint (LCP):** Measures the time it takes for the largest content element (like an image or text block) to load on the page.

- **Interaction To Next Paint (INP):** Measures a page's overall responsiveness by tracking the time from when a user interacts with the page (e.g., clicks a button) to when the browser paints the next frame.
- **Cumulative Layout Shift (CLS):** Measures visual stability by quantifying unexpected shifts in the page layout when the display dimensions are changed.

Keeping layouts stable and interactions smooth helps to avoid being penalized by search engines.

Another metric search engines use to score user experience is whether links work. Search engines like Google and Bing will take note if clicking on a link results in a 4XX (page not found) error.

The best way to avoid penalizations and maintain technical integrity is to regularly audit your website. Tools like Screaming Frog can audit your site for broken links, incorrect canonicals, or blocked resources.

CORE METRICS FOR AEO PERFORMANCE

To measure your AEO impact, focus on metrics that reflect both human and machine selection:

- **Featured Snippet Wins:** Track how often your content appears in position zero. Use Google Search Console and third-party SERP tracking tools.
- **PAA Placements:** Monitor the number and variety of People Also Ask questions your content answers. Tools like SEMrush, Ahrefs, and AlsoAsked reveal these opportunities.
- **Voice Search Results:** Test your answers on devices and simulators (Google Assistant, Siri, Alexa, Jetson.ai). Note which questions trigger your content as the spoken answer.
- **AI Citations and Mentions:** Use prompt testing in ChatGPT, Bing Copilot, Perplexity, and other generative platforms. Watch for direct citations, paraphrased answers, and inclusion in summaries.

- **Zero-Click and Branded Traffic:** Analyze analytics for increases in impressions without clicks, branded search volume, and referral sources from AI tools.
- **Local Pack and Map Results:** For local AEO, track your Google Business Profile visibility, map pack appearances, and consistency across directories.
- **Page Experience Metrics:** Optimize for Largest Contentful Paint (LCP), Interaction To Next Paint (INP), and Cumulative Layout Shift (CLS).

TOOLS FOR CROSS-PLATFORM MONITORING

No single tool covers every channel. Combine standard SEO platforms with specialized AEO and AI-tracking solutions:

- **Google Search Console:** Essential for tracking impressions, clicks, rich results, and mobile usability.
- **Bing Webmaster Tools:** Complements Google data and helps track Bing-specific features.
- **SEMrush, Ahrefs, Moz:** Benchmark keyword rankings, snippet placements, PAA appearances, and competitor performance.
- **AlsoAsked, Answer the Public:** Surface real user questions and track your answers in PAA and related queries.
- **Voice Search Simulators:** Test your content's spoken responses with Jetson.ai, Seedtag Voice, and Google Assistant Simulator.
- **Brand Monitoring Tools:** Use platforms like Brand24, Mention, or Google Alerts to track citations and mentions in AI-generated content and summaries.
- **AI Prompt Testing:** Regularly prompt ChatGPT, Bing Copilot, Perplexity, and other LLMs with your target queries to see if your content is retrieved or cited.

Interpreting and Acting on Data

Raw data is only the start, you must actually use the information.

Look for trends and patterns:

- Are you gaining more featured snippets or PAA placements over time?
- Do you see an increase in zero-click or branded traffic?
- Are your answers being cited in AI outputs or voice search?
- Where are you losing visibility, and what changed?

Set clear benchmarks for each channel. Compare your performance against competitors. Identify gaps—questions you're not answering, platforms where you're absent, or technical issues that block visibility.

When you spot a drop in performance, diagnose the cause. Did a recent algorithm update change snippet selection? Has your schema markup broken? Is your content outdated or less relevant? Use these insights to refine your strategy, update assets, and close gaps.

Continuous Improvement and Scheduled Audits

AEO monitoring is not a one-off project. Build a review process:

- **Monthly:** Track featured snippet wins, PAA placements, and voice search results. Run prompt tests in major LLMs.
- **Quarterly:** Audit schema markup, technical SEO, and accessibility. Review Google Business Profile and local listings for accuracy.
- **Annually:** Quantify your authority footprint, citation frequency, and inclusion in training datasets or knowledge panels.

Document your findings. Share insights with your team. Use what you learn to inform your next round of content updates, technical fixes, and strategic pivots.

KEY TAKEAWAYS

- Monitoring AEO performance means tracking visibility, citations, and engagement across search, voice, and AI-driven platforms—not just rankings.
- Use a mix of analytics, SERP tools, brand monitoring, and AI prompt testing to get a complete view of your presence.

CHAPTER 22

Debugging Ranking Drops and Indexing Issues

DIAGNOSING THE PROBLEM

Ranking drops and indexing issues can happen suddenly or creep in over time. They threaten your visibility across search engines, answer engines, and AI platforms. When your content stops appearing where it should, you lose traffic, citations, and authority. **The key is to diagnose the root cause quickly, fix it thoroughly, and monitor for recurrence.**

You'll notice ranking drops in your analytics—traffic declines, lost featured snippets, or fewer People Also Ask appearances. Indexing issues show up as pages missing from search results, flagged in Google Search Console, or not being cited by AI systems. Sometimes the cause is technical. Other times, it's content quality, algorithm changes, or external signals.

Effective debugging is systematic. Don't guess. Use data, tools, and a step-by-step approach to pinpoint and resolve the issue.

PRO TIPS AND COMMON PITFALLS

- **Don't panic and make sweeping changes without diagnosis.** Isolate the problem before acting.
- **Document every change.** Keep a log of what you fixed and when, to correlate with recovery.
- **Test on multiple devices and platforms.** Sometimes issues are browser- or device-specific.
- **Stay proactive.** Schedule regular technical audits, content reviews, and backlink checks to catch problems early.

Step 1: Confirm and Isolate the Drop

First, confirm that the drop is real and not a temporary fluctuation. Compare analytics data across platforms (Google Search Console, Bing Webmaster Tools, third-party SERP trackers).

Look for:

- **Sudden or gradual declines** in impressions, clicks, or rankings
- **Loss of featured snippets**, PAA, or voice search results
- **Pages dropping out** of the index or not being cited by LLMs

Isolate which URLs, topics, or types of queries are affected. Is the problem sitewide, or limited to a section or page? Are you losing ground in organic search, AI citations, or both?

Step 2: Check for Technical Issues

Technical errors are a common culprit.

Audit your site for:

- **Crawlability problems:** Check robots.txt, noindex tags, canonical tags, and sitemap accuracy. Use Google Search Console's Coverage report to spot blocked or excluded pages.
- **Broken links and redirects:** Scan for 404 errors, soft 404s, incorrect redirects, or broken internal links. Tools like Screaming Frog or Sitebulb can help.
- **Server errors:** Look for 5xx errors, downtime, or slow response times. These can cause pages to drop from the index.
- **Mobile and Core Web Vitals:** Ensure your site is mobile-friendly and meets speed, stability, and usability benchmarks. Poor performance can lower rankings.

- **Schema and structured data:** Validate your markup with Google's Rich Results Test. Broken or missing schema can make your content invisible to answer engines and AI.

Fix technical issues before moving on. Many ranking drops are resolved by restoring crawlability, fixing errors, and ensuring your site is accessible to both users and machines.

STEP 3: REVIEW CONTENT QUALITY AND RELEVANCE

If technical health checks out, evaluate your content.

Ask these questions about your content:

- **Is it up to date?** Outdated information loses rankings and citations, especially for time-sensitive topics.
- **Does it match search intent?** Algorithms now prioritize relevance and clarity. Make sure your answers are direct, structured, and fulfill user needs.
- **Is it authoritative?** Thin, duplicate, or low-quality content is deprioritized. Ensure your pages demonstrate E-E-A-T (Experience, Expertise, Authoritativeness, Trustworthiness).
- **Are you over-optimized?** Excessive keyword stuffing, manipulative schema, or unnatural internal linking can trigger penalties or demotions.

Refresh, expand, or consolidate content where needed. Remove or update anything that no longer serves your audience or aligns with current best practices.

STEP 4: ANALYZE EXTERNAL SIGNALS

If your content is fine, look at external factors that can impact your rankings and indexing.

Check:

- **Backlinks:** Loss of high-quality backlinks or a spike in spammy links can cause drops. Use a backlink analysis tool to review your profile and disavow harmful links if necessary.
- **Citations and reviews:** For local and entity-based queries, inconsistent NAP data or negative reviews can hurt visibility. Ensure all business listings are accurate and up to date.
- **Competitor changes:** Sometimes your drop is due to others improving. Analyze who is now ranking or being cited in your place, and study what they're doing differently.

STEP 5: CHECK FOR ALGORITHM UPDATES AND MANUAL ACTIONS

Algorithm updates can shift ranking factors overnight. **Make sure to stay informed:**

- Review Google's Search Status Dashboard and industry news for recent updates.
- Check Google Search Console for manual actions, security issues, or policy violations.
- If you've been penalized, follow Google's guidance to resolve the issue and request reconsideration.

STEP 6: REINDEX AND MONITOR

Once you've made fixes, request reindexing in Google Search Console or Bing Webmaster Tools. **Submit updated sitemaps**. Use IndexNow for instant updates where supported.

Monitor your analytics and rankings closely over the next few weeks. Look for signs of recovery—returning impressions, traffic, snippet wins, or AI citations. If the issue persists, repeat your audit or seek expert help.

Key Takeaways

- Ranking drops and indexing issues can stem from technical, content, external, or algorithmic causes.
- Use a structured, step-by-step process to diagnose and resolve issues—don't rely on guesswork.

CHAPTER 23

Iteration and Lifecycle Management of Assets

Why Lifecycle Management Matters

Answer Engine Optimization (AEO) isn't a one-time project. The digital landscape, user intent, and AI retrieval logic all evolve. Content that wins today can fade tomorrow. **To stay visible and trusted, you need a process for continuous improvement—one that treats your AEO assets as living resources, not static pages.**

Lifecycle management is about more than updates. It's about strategic iteration, regular audits, and knowing when to refresh, repurpose, or retire assets. This approach keeps your content aligned with current best practices, search algorithms, and user expectations.

You'll notice that top-performing sites rarely let content stagnate. They review, optimize, and adapt. This is how you build durable authority and maintain your place as the answer—no matter how the ecosystem shifts.

Building an Iterative Content Process

Make sure your best pages keep evolving with the questions, competitors, and formats in your space.

Start with a schedule. Set regular intervals for reviewing your pillar pages, cluster content, and supporting assets. For high-value topics, monthly or quarterly reviews are ideal. For evergreen content, biannual or annual check-ins may suffice.

During each review, ask:

- Is the information still accurate and complete?

- Has user intent shifted, based on new queries or analytics?
- Are there new competitors or formats dominating the answer space?
- Do schema markup, metadata, and internal links still align with best practices?

Document your findings. Use a content calendar or asset tracker to log updates, flag issues, and plan future improvements. This record helps you correlate changes with performance shifts and avoid repeating mistakes.

Refreshing and Expanding Existing Assets

When an asset starts to slip in rankings or loses citations, don't scrap it—**refresh it.** Update statistics, add new research, and address emerging questions. Expand sections that are thin or outdated. Remove obsolete references or broken links.

Look for opportunities to add value:

- **Include** new FAQs based on People Also Ask trends.
- **Add** case studies, testimonials, or real-world examples.
- **Embed** updated videos, infographics, or interactive elements.
- **Refine** headings and summaries for clarity and scannability.

If a page is too broad, consider splitting it into more focused subtopics. If it's too narrow, merge it with related content to create a more authoritative resource.

Repurposing and Distributing Content

Great AEO assets can be repurposed across formats and channels. Turn a high-performing guide into a downloadable PDF, a series of blog posts, or a video tutorial. Summarize key sections for social media or email newsletters.

Distribute refreshed assets to platforms where AI and users discover answers: Internet Archive, arXiv, YouTube, Google Books, and trusted directories. Update schema and metadata for each new format.

Cross-link between versions. For example, link your PDF to the original web page, or embed your video in the related article. This reinforces authority and helps both users and machines connect the dots.

Retiring or Consolidating Underperforming Assets

Not every page deserves to live forever. If an asset consistently underperforms, is no longer relevant, or risks diluting your authority, consider retiring or consolidating it.

Redirect retired URLs to the most relevant, updated resource. Merge thin or overlapping pages into a single, comprehensive guide. Remove outdated or low-quality content that no longer aligns with your brand or expertise.

Document these changes. Keep a log of retired assets and the rationale behind each decision. This helps maintain a clean, authoritative site and avoids confusion for users and AI systems.

Key Takeaways

- Lifecycle management is essential for long-term AEO success. Treat your content as a living asset, not a static deliverable.
- Regularly review, refresh, and expand your assets to maintain authority and visibility.

CHAPTER 24

Ethical Considerations and Transparency in the AI Age

THE NEW STANDARD FOR DIGITAL ETHICS

As AI-driven search and answer engines become the norm, ethical considerations are no longer optional. The choices you make—what you publish, how you structure it, and how you represent expertise—directly shape the information ecosystem.

In the answer economy, trust and transparency are not just compliance requirements. They are the foundation of authority and lasting visibility.

You'll notice that users, regulators, and AI platforms all demand more: clearer sourcing, honest representation, and accountability for the information you provide. The risks of misinformation, manipulation, or hidden bias are higher than ever. At the same time, the rewards for ethical leadership—being cited, trusted, and chosen—are substantial.

TRUTHFULNESS, ATTRIBUTION, AND FACT-CHECKING

Every answer you publish may be surfaced, summarized, or cited by AI—sometimes out of context or at scale. Accuracy is paramount.

Fact-check every claim, statistic, and recommendation. Use reputable sources and cite them clearly, both for human readers and for machines.

Attribution is not just about avoiding plagiarism. It's about giving credit, building trust, and allowing users (and AI) to verify your claims. When you reference studies, expert opinions, or third-party data, link directly to the original source.

If you use generative AI for drafting or research, disclose this in your methodology or acknowledgments.

Transparency in sourcing helps users understand where information comes from and gives AI models better signals for citation and retrieval. It also protects your reputation if a claim is later challenged or revised.

Disclosures, Authorship, and AI Involvement

Clear authorship is essential. Every page, article, or guide should identify its creator, their credentials, and their role in the content's production. This is especially important as AI-generated and AI-assisted content proliferates. **If AI tools contributed to research, drafting, or editing, disclose this plainly**—ideally in a dedicated section or footnote.

Disclosures build trust. They show you value honesty over expedience. They also help AI systems and knowledge graphs correctly attribute expertise and avoid confusion with similarly named entities.

For organizations, clarify editorial oversight. Who reviewed the content? What standards were applied? This level of transparency is increasingly expected, especially for health, finance, legal, or other sensitive topics.

Avoiding Manipulation and Bias

AI systems are designed to surface the best answers, but **they are not immune to manipulation**. Practices like keyword stuffing, cloaking, or artificially inflating reviews are not just unethical—they can lead to penalties, deindexing, or reputational harm.

Be vigilant about bias. Review your content for unintentional slant, exclusion, or stereotyping. Strive for balanced coverage, especially when addressing controversial or evolving topics. If your content represents an opinion, label it as such and present alternative viewpoints when appropriate.

Regular audits help catch issues before they become problems. Use both human review and AI tools to scan for misleading phrasing, outdated information, or gaps in representation.

Privacy, Consent, and User Data

Respecting user privacy is a core ethical principle. If your content collects, processes, or references user data, be clear about how that data is used.

Follow all relevant regulations (GDPR, CCPA, etc.) and avoid unnecessary data collection.

Obtain consent for testimonials, case studies, or any user-generated content you publish.

Anonymize sensitive details where possible. Transparency about data practices not only builds trust with users but also aligns with the standards AI platforms use to evaluate trustworthy sources.

The Role of Continuous Improvement

Ethical standards are not static. As AI and search platforms evolve, so do expectations around transparency, accuracy, and fairness. Build regular ethical reviews into your content lifecycle. Encourage feedback from users and peers. Stay informed about changes in policy, law, and best practices.

When you make a correction or update, document it. Let users know when and why information changed. This openness signals integrity and helps AI systems maintain accurate citations.

Key Takeaways

- Ethical AEO is built on accuracy, clear attribution, and honest disclosures.
- Transparency about authorship, AI involvement, and editorial standards builds trust with both users and machines.

CHAPTER 25

The Future of AEO—Strategic Foresight

NAVIGATING A RAPIDLY CHANGING LANDSCAPE

The answer economy is still in its early days. Search, content, and AI are evolving at a pace that challenges even the most adaptive organizations. What works now—structured data, pillar content, entity optimization—will remain essential, but the tactics and tools will keep shifting.

Your success in AEO depends on anticipating these changes and building a strategy that's flexible, resilient, and always learning.

You'll notice that user expectations are rising. **People want answers, not just information.** They expect accuracy, transparency, and context—delivered instantly, whether they're searching on a screen, speaking to a voice assistant, or interacting with a chatbot. At the same time, AI models are becoming more sophisticated, drawing from broader datasets and increasingly prioritizing trust, clarity, and authority.

EMERGING TRENDS IN ANSWER ENGINE OPTIMIZATION

Several trends are already shaping the future of AEO.

Generative AI is integrating deeper into search platforms, producing summaries and responses that blend live retrieval with trained knowledge.

Retrieval-augmented generation (RAG) is becoming the norm, making it critical for content to be both retrievable and citable.

Voice and multimodal search are expanding. Users are searching with images, video, and spoken queries—sometimes all at once. This means your optimization efforts must extend beyond text, ensuring every asset is structured, accessible, and easy to extract.

Entity-first retrieval is gaining ground. Search engines and LLMs are moving from page-based indexing to entity-based mapping. Being recognized as a trusted entity—person, organization, or product—will be a prerequisite for citation and selection.

Expect compliance and ethical standards to tighten. Transparency about authorship, AI involvement, and sourcing will become mandatory. Content that is misleading, manipulative, or opaque will be filtered out by both algorithms and regulatory frameworks.

Strategic Foresight: How to Prepare

To future-proof your AEO strategy, focus on adaptability and continuous improvement.

Build processes for regular content audits, technical updates, and ethical reviews. Stay informed about changes in search algorithms, AI training practices, and user behavior.

Invest in structured data and semantic clarity. These are not passing trends—they are the foundation for discoverability in any AI-driven environment. Expand your content formats, embracing video, audio, and interactive assets. Make sure every asset is accessible, well-labeled, and easy for machines to parse.

Deepen your entity footprint. Claim and maintain knowledge panels, Wikidata entries, and authoritative profiles. Link your content and identity across platforms. The more consistently you present yourself as a trusted entity, the more likely you are to be surfaced and cited by future AI systems.

Monitor new channels and interfaces. AI-powered search will not be limited to Google or Bing. Prepare for discovery in chatbots, voice assistants, and new generative interfaces as they emerge. Test your content in these spaces and adapt quickly.

EMBRACING A CULTURE OF EXPERIMENTATION

The most successful brands and experts in the answer economy will be those who experiment, measure, and iterate.

Treat every update as a learning opportunity. Use analytics, manual testing, and prompt engineering to see how your content performs in real-world AI and search scenarios.

Encourage collaboration across teams—content, technical, legal, and ethical. The future of AEO is multidisciplinary. It rewards those who break down silos and approach optimization as a holistic, ongoing process.

Stay open to new technologies and methodologies. As AI evolves, new tools for monitoring, optimizing, and measuring visibility will emerge. Don't be afraid to test them, but always ground your strategy in clarity, authority, and trust.

KEY TAKEAWAYS

- The future of AEO is dynamic. Expect rapid changes in search technology, user behavior, and compliance standards.
- Structured data, entity optimization, and ethical transparency will remain foundational, even as tactics evolve.

Chapter 26

The LLM SEO Toolkit—Tools to Operationalize Your Strategy

Why Tools Matter for LLM SEO

Optimizing for large language models (LLMs) and answer engines requires more than intuition.

The complexity of modern search, AI-driven retrieval, and entity-based visibility means you need the right toolkit to operationalize your strategy. Tools help you uncover opportunities, monitor performance, automate structured data, and validate that your content is not just visible, but chosen and cited.

You'll notice that top performers in the answer economy use a blend of traditional SEO platforms and new AI-specific solutions. The right mix lets you move faster, spot gaps before they cost you visibility, and adapt as new channels and formats emerge.

Pro Tips for Tool Selection

- **Choose tools that integrate** with your existing stack and support your workflow.
- **Prioritize platforms that update frequently** to reflect changes in search and AI algorithms.
- **Test new tools on a small scale** before rolling them out across your entire strategy.
- **Don't rely solely on automation**—manual review and prompt testing remain essential for real-world visibility.

Discovery and Opportunity Mapping

Scouting out AEO opportunities starts with research.

For understanding keyword demand, tracking rankings, and benchmarking against competitors, tools like SEMrush, Ahrefs, and Moz remain essential. They now offer features for tracking featured snippets, People Also Ask (PAA) placements, and zero-click results—key signals for AEO.

For mapping user questions and intent clusters, use AlsoAsked, Answer the Public, and Frase. These platforms surface real-world queries, reveal intent patterns, and help you identify gaps in your current coverage. MarketMuse and Clearscope go further, analyzing semantic fields and suggesting subtopics to deepen your authority.

For local and entity-based discovery, BrightLocal and Moz Local audit your NAP consistency and local search presence, ensuring your business is surfaced for "near me" and voice queries.

Content Creation and Optimization

Drafting and refining AEO assets is faster and more precise with AI-powered writing assistants.

For help with initial drafting, Jasper, Writesonic, and Claude help generate outlines, draft answer-first paragraphs, and repurpose content into FAQs or snippet-ready blocks. Use these tools to accelerate production, but always review for accuracy, tone, and authority.

For semantic optimization, SurferSEO, Clearscope, and Frase analyze your drafts for topic coverage, entity inclusion, and semantic alignment. They highlight missing concepts and suggest improvements to match what LLMs and answer engines expect.

For proofreading and editing, Grammarly and Hemingway Editor ensure your writing is clear, concise, and free of ambiguity—critical for both human readers and AI retrieval.

STRUCTURED DATA AND TECHNICAL VALIDATION

Schema markup is the backbone of machine readability, it makes your page easier to understand and selectively process by AI.

Add schema when you create your content. Merkle's Schema Markup Generator and Schema.dev allow you to build and validate JSON-LD for FAQ, HowTo, Article, LocalBusiness, VideoObject, and Book schemas.

IndexNow and XML sitemap generators help you push updates to search engines and AI crawlers instantly, reducing lag between publishing and discovery.

Google provides Schema validation tools. Use Google's Rich Results Test and Schema Markup Validator to catch errors before they impact visibility.

For technical audits, Screaming Frog and Sitebulb crawl your site for broken links, crawlability issues, and schema implementation. Google Search Console and Bing Webmaster Tools provide direct feedback on indexing, rich results, and mobile usability.

AI VISIBILITY AND PROMPT TESTING

Visibility in LLMs and generative search platforms is best measured through hands-on testing.

Regularly prompt AI search tools with your target queries. Use ChatGPT, Bing Copilot, Perplexity, and other popular tools on the market. Note if your content is cited, paraphrased, or ignored. Adjust your structure, phrasing, or schema based on these results.

Listen to how answers are delivered by voice interfaces. Jetson.ai, Seedtag Voice, and Google Assistant Simulator let you test how your answers sound in conversational interfaces. This is crucial for optimizing spoken responses and local queries.

For brand and citation monitoring, use Brand24, Mention, or Google Alerts to track references to your content in AI-generated summaries, reviews, and knowledge panels.

Knowledge Distribution and Entity Management

In an AI-first ecosystem, it's not enough to publish content—you also have to teach machines who you are and what you've created, by managing your entities and distributing your work through the channels LLMs and knowledge graphs actually trust.

To maximize your presence in LLM training data and knowledge graphs, use tools like Wikidata, Google Knowledge Panel Manager, and ISNI (International Standard Name Identifier) to claim and maintain your entity profiles.

For books and long-form content, Bowker and Nielsen handle ISBN registration and metadata management. Google Books Partner Center ensures your book metadata is accurate and discoverable.

For distributing open-access research and technical assets, arXiv, SSRN, and Internet Archive are essential. These platforms are routinely crawled by LLMs and increase your odds of training data inclusion.

Workflow and Collaboration

Answer Engine Optimization works best as a team sport: the more you centralize your tasks, timelines, and automation in shared workflows, the easier it is to keep everyone shipping aligned, consistent, high-quality content.

Keep your process organized with project management tools like Trello, Asana, or Notion. Use shared content calendars and asset trackers to log updates, schedule audits, and coordinate across teams.

Automate routine checks with Zapier or Make (formerly Integromat), connecting analytics, monitoring, and publishing workflows. If you'd like to get more advanced, you can try using n8n or Flowise.

KEY TAKEAWAYS

- The LLM SEO toolkit blends classic SEO platforms with AI-specific discovery, writing, and validation tools.
- Use research, semantic optimization, and structured data generators to build assets that answer engines and LLMs can select, cite, and trust.

CHAPTER 27

Final Checklist and AEO Maturity Model

The Routine Tasks of AEO

Part IV brought your AEO strategy full circle, focusing on what it takes to sustain, measure, and future-proof your visibility in the answer economy.

Here, you learned that **success is not just about initial optimization, but about ongoing vigilance, ethical leadership, and adaptability.**

This section guided you through the realities of monitoring your performance across platforms, debugging ranking drops, and managing the lifecycle of your content assets. You explored how to build a process for continuous improvement—auditing, updating, and repurposing your work to stay ahead of shifting algorithms and answer interfaces.

We also examined the ethical responsibilities that come with authority in the AI age, emphasizing transparency, accuracy, and respect for user trust. Looking forward, you gained strategic foresight to anticipate changes in search and AI, ensuring your expertise remains relevant and chosen as new channels and technologies emerge.

Finally, you were equipped with a practical toolkit to operationalize your AEO efforts, benchmark your progress, and set your sights on the next level of leadership. The result is a resilient, adaptive approach—one that keeps your content visible, cited, and trusted in a landscape that never stands still.

THE AEO MATURITY MODEL

Use this maturity model to assess where you stand and identify what to prioritize next. Each stage builds on the last, moving from basic compliance to true answer leadership.

Level	Description	Focus Areas
1. Reactive	Basic SEO in place, but no AEO structure. Content is unstructured and rarely cited by AI.	Fix technical errors, add schema, clarify metadata.
2. Structured	Pillar pages and supporting content exist. Schema markup is implemented. Internal linking is logical.	Audit for gaps, optimize for snippets and PAA, expand entity profiles.
3. Optimized	Content is answer-ready, cited by AI, and visible in featured snippets and voice. Authority signals are consistent.	Monitor AI citations, refresh content, build external authority.
4. Strategic	Cross-channel optimization is routine. Assets are distributed for LLM training and entity recognition. Iteration is scheduled.	Publish open-access works, manage knowledge panels, test AI prompts.
5. Leading	Brand is a trusted entity, regularly cited by LLMs and answer engines. Continuous improvement is embedded in culture.	Experiment with new formats, collaborate across teams, lead in ethical transparency.

AEO SUSTAINABILITY & GROWTH CHECKLIST

Measurement and Monitoring

- ❐ I have defined clear benchmarks for AEO performance across all major platforms (search, voice, AI, local, video, and books).
- ❐ I use a mix of analytics, SERP tools, brand monitoring, and prompt testing to track visibility, citations, and answer placements.
- ❐ I analyze trends in zero-click traffic, branded queries, and cross-channel mentions to understand my evolving authority.

Debugging and Technical Resilience

- ❐ I have a process for quickly diagnosing and resolving ranking drops or indexing issues, using tools like Google Search Console, Screaming Frog, and schema validators.
- ❐ I routinely audit for crawlability, broken links, schema errors, and mobile performance.
- ❐ I document every technical or content change and monitor for recovery or further issues after fixes.

Lifecycle Management and Iteration

- ❐ I schedule regular audits of pillar content, supporting assets, and technical SEO.
- ❐ I refresh, expand, or consolidate content based on performance, user feedback, and changing search/AI trends.
- ❐ I repurpose high-performing content into new formats (PDFs, videos, social posts) and distribute across relevant channels.

Ethical Standards and Transparency

- ❐ I fact-check all claims, cite reputable sources, and provide clear attribution for research and data.
- ❐ I disclose authorship, editorial oversight, and any AI involvement in content creation.
- ❐ I obtain consent and anonymize user data where required, and comply with relevant privacy regulations.

Strategic Foresight and Adaptability

- ❒ I stay informed about changes in search algorithms, AI retrieval and training practices, and compliance requirements.
- ❒ I monitor emerging channels (chatbots, AR, new AI platforms) and test my content in these environments.
- ❒ I experiment with new tools, formats, and workflows, documenting results to inform future strategy.

CONGRATULATIONS ON REACHING THE END OF THIS GUIDE.

Use what you've learned here as a foundation, keep testing and refining your approach, and you'll continue to set the standard for visibility and impact in the age of answer.

Glossary

AAEO (Agentic Answer Engine Optimization)

A variation of AEO that focuses on how autonomous agents (for example, AI assistants that browse, compare, and act on behalf of users) discover, evaluate, and reuse your content as part of multi-step tasks.

AEO (Answer Engine Optimization)

The discipline of making your content the answer that gets selected, cited, or spoken by search engines, AI models, and answer interfaces—not just the page that ranks. It focuses on being included in training data, retrieved in context, and chosen in the final output.

AEO Maturity Model

A staged framework in Part IV that describes how far along a brand or organization is in building a durable, repeatable AEO practice—from ad-hoc experiments to fully integrated, cross-channel visibility systems.

Answer card / Knowledge card

A compact, structured answer format generated by Google or an AI engine that displays facts, definitions, or summaries without requiring a click. Often powered by structured data or strong entity signals.

Answer engine

Any system that directly returns synthesized answers instead of (or in addition to) a list of links. Examples include Google's Search Generative Experience (SGE), Bing Copilot, ChatGPT, Perplexity, voice assistants, and other generative interfaces.

Answer layer

The layer where decisions are made about which content is surfaced as the "final answer." It includes featured snippets, "People Also Ask," SGE summaries, voice responses, and AI chat outputs.

Answer selection

The process by which an answer engine chooses which passage, paragraph, or source to surface. Distinct from ranking. Selection is driven by clarity, structure, trust signals, and retrieval quality.

Asset lifecycle

The end-to-end life of a content asset—from initial creation and launch through updates, repurposing, consolidation, and eventual retirement (or archival) as part of ongoing AEO maintenance.

Authoritativeness (in E-E-A-T)

The degree to which a person, brand, or site is recognized as a trusted source on a specific topic—often signaled by citations, high-quality backlinks, strong entity presence, and consistent expert coverage.

Backlink

A link from one site to another. In traditional SEO, backlinks act as "votes of confidence." For AEO, they still matter, but mainly as part of broader authority and entity signals rather than as the sole ranking factor.

Brand/entity footprint

The totality of how your brand, organization, or products appear across the web: websites, profiles, directories, knowledge panels, Wikidata, books, and more. A strong, consistent footprint helps AI systems "connect the dots" and trust your content.

Canonical URL

The preferred URL for a piece of content, used to prevent duplicate content issues.

Channel

A specific discovery or delivery surface where content can appear, such as web search, local search, marketplaces, YouTube, voice assistants, in-app search, or AI chat interfaces.

Citation (AI/answer context)

A reference within an AI or search result pointing back to your content as the source of an answer. In the answer economy, citations are the new “clicks”—evidence that your content is being trusted and reused.

Cluster (topic cluster / content cluster)

A group of related content pieces that all support a central pillar topic. Clusters typically include FAQs, supporting articles, comparisons, and how-to content that deepen coverage and help search engines and LLMs understand your topical authority

Content freshness / recency

A relevance signal reflecting how recently a page or asset was updated. Increasingly important for AI systems that prioritize up-to-date sources when generating answers.

Context window

The maximum amount of text (tokens) an LLM can use at one time for reasoning, retrieval, and generation. Answer selection depends on whether your content fits cleanly into the model’s context window during retrieval.

Conversational query

A natural-language question (for example, "What's the best way to...?") rather than a short keyword string. Conversational queries are common in voice search, chat interfaces, and modern search behavior.

Crawling

The process by which search engines and some AI systems discover new and updated pages using automated bots ("crawlers" or "spiders"). If a page cannot be crawled, it cannot be indexed or surfaced.

Content ecosystem

The interconnected network of pages, assets, and channels that together represent your brand's knowledge on a topic. A healthy ecosystem includes pillar pages, clusters, FAQs, structured data, and consistent entity signals across platforms.

Content decay

The gradual decline in performance of content due to outdated information, changing SERPs, or newer competitors. A focus area in AEO maintenance cycles.

Content intent

What your content is designed to accomplish for the user (inform, compare, convince, convert) and how that maps to search intent (informational, commercial, transactional, navigational).

Data source (LLM context)

Any text, document, or asset a model uses at inference time, including indexed pages, retrieved passages, structured data, and user-provided content.

Deep content

Comprehensive, structured coverage of a topic that anticipates follow-up questions, includes supporting detail, and can be selectively reused by answer engines. Often aligned with pillar content or major guides.

Document chunking

Breaking large documents into smaller segments (embeddings) so AI systems can retrieve specific parts instead of entire pages. Essential for RAG-friendly formatting.

Disambiguation (entity context)

The process by which search engines and AI decide which entity you represent when multiple entities share similar names (e.g., "Apple," "Mercury," "Atlas"). Critical for brands and personal names.

E-E-A-T (Experience, Expertise, Authoritativeness, Trustworthiness)

A framework used by search engines and increasingly by AI systems to assess whether content should be trusted and surfaced. AEO treats E-E-A-T as the backbone of both human credibility and machine selection.

Embedding

A numerical representation of text that captures its meaning and context. LLMs use embeddings to compare concepts, find similar content, and retrieve relevant passages during question answering.

Entity

A distinct "thing" that can be recognized by machines: a person, organization, product, place, concept, or work (like a book). Entities live in knowledge graphs and knowledge bases; strengthening your entity presence improves your odds of being cited.

Entity optimization

The practice of making sure your entities (brand, products, key people, books) are clearly defined and connected across the web. This includes knowledge panels, Wikidata entries, consistent naming, structured data, and citations in trusted sources.

Featured snippet

A highlighted answer box at the top of Google's results that directly answers a query, often pulling a short paragraph, list, or table from a page. A core surface where AEO shows up in traditional SERPs.

GEO (Generative Engine Optimization)

A related term to AEO that emphasizes optimizing for generative engines (like SGE, ChatGPT plugins, or Perplexity) that create synthesized responses using AI.

Hallucination (AI context)

A fabricated or incorrect statement generated by an AI model. Strong AEO reduces hallucination risk by providing clear, structured, trustworthy content for models to select and cite.

HTML semantics

The use of proper headings, lists, tables, and markup so crawlers and AI systems can accurately parse and understand content structure.

Indexing

The process of storing crawled pages in a searchable database. If your page is not indexed, it cannot appear in search or be used in some retrieval flows.

Information gain

The degree to which content adds new, unique, or more useful information compared to other online sources. A known factor in snippet selection.

Intent (search intent / user intent)

The underlying goal behind a query: what the user is actually trying to achieve. Often categorized as informational, commercial, transactional, or navigational. AEO aligns content and formatting to the intent, not just the keywords.

JSON-LD

A common format for implementing schema markup on a web page, typically embedded in a <script type="application/ld+json"> block. Preferred by Google for structured data implementation.

Knowledge base

A structured repository (local or cloud) used by AI or internal systems to store canonical content, reference answers, definitions, or product data. Often integrated into enterprise RAG systems.

Knowledge graph

A structured network of entities and their relationships, maintained by systems like Google, Microsoft, and Wikidata. Knowledge graphs help answer engines understand how people, places, brands, and concepts connect.

Knowledge panel

An information box (usually on the right-hand side of Google desktop results or at the top on mobile) that aggregates key details about an entity. A strong knowledge panel is a signal of recognized authority.

LLM (Large Language Model)

An AI model trained on large text datasets to generate and interpret language. Models like GPT, Gemini, Claude, and others underpin many answer engines and chat interfaces.

LLM SEO

The strategic layer of optimizing for large language models across training, retrieval, and output—ensuring your content is memorable, retrievable, and cite-worthy in AI systems.

Local AEO

Answer Engine Optimization focused on "near me" and location-dependent queries. It relies heavily on NAP consistency, localBusiness schema, reviews, local packs, and voice search behavior.

Metadata (SEO/AEO context)

Information about a content asset—titles, descriptions, publication dates, schema, author information—that helps machines understand and evaluate the asset.

NAP (Name, Address, Phone)

Core business identity data for local search. Consistent NAP across your website, Google Business Profile, directories, and citations is crucial for local visibility and answer selection.

People Also Ask (PAA)

Expandable question boxes in Google that list related queries and short answers. A key AEO surface for capturing mid-funnel and follow-up questions.

Pillar content (pillar page / pillar asset)

A comprehensive, authoritative page or asset that covers a core topic in depth and acts as the anchor for supporting content. Pillars are structured for retrieval, internal linking, and snippet/answer reuse.

Prompt

The text or instructions you give to an AI model. Prompt design affects how the model interprets your intent, what it retrieves, and how it cites or summarizes information.

Prompt engineering (for AEO)

The practice of systematically testing and refining prompts to see how answer engines surface your content—and then adjusting your content and structure so it's more likely to be selected or cited.

RAG (Retrieval-Augmented Generation)

A method where an LLM retrieves relevant documents or passages at query time and uses them to inform its response. For AEO, RAG makes formatting, structure, and machine readability even more important.

Ranking

The order in which search results, snippets, or answer surfaces appear. Traditional SEO focuses heavily on ranking in SERPs; AEO focuses on selection within answer layers, which may or may not resemble standard rankings.

Retrieval (in LLMs and AEO)

The process by which an AI model pulls in relevant passages, documents, or embeddings to answer a query. Retrieval determines what the model can use before selection determines what it does use.

Schema markup (structured data)

Machine-readable annotations (usually JSON-LD) added to pages to help search engines and AI understand what the content represents—articles, FAQs, products, books, local businesses, how-tos, and more.

Schema type

The specific structured data category used for markup (Article, FAQPage, Product, Course, LocalBusiness, etc.).

Search Generative Experience (SGE)

Google's AI-driven interface that generates conversational summaries and answer overviews at the top of search results, often with citations to sources. A primary answer engine surface for AEO.

Search intent overlap

The degree to which different user intents (informational, commercial, transactional) cluster together around a topic. High overlap means your content must serve multiple intents simultaneously to win answers.

Semantic density

How tightly a passage conveys meaning with minimal filler. High semantic density improves snippet extraction and LLM retrieval accuracy.

Semantic search / semantic understanding

Search and AI behavior that focuses on meaning and context instead of exact keyword matches. Semantic systems use embeddings and entity graphs to understand what a query and document are "about."

SERP (Search Engine Results Page)

The page returned by a search engine in response to a query. It may contain ads, organic results, featured snippets, PAA, local packs, images, videos, and generative answer panels.

Snippet-ready content

Content deliberately formatted so that a short, self-contained piece (paragraph, list, table, or step-by-step) can be extracted and reused as a snippet or answer, without additional context.

Structured data

Any information on a page that is labeled and formatted so that machines can reliably parse it. On the web, this usually refers to schema markup, but can also include well-structured tables, lists, and consistent HTML headings.

Thin content

Content with little value, insufficient depth, or poor structure. Historically penalized by search engines and now deprioritized by AI systems for retrieval and answer selection.

Token

A small unit of text (a word or sub-word) used internally by LLMs. Models process tokens rather than full words; tokenization shapes how they "see" and reconstruct language.

Topical authority

The collective signal that you are an expert source on a subject, determined by coverage depth, consistency, entity presence, and reinforced internally through clusters and externally through citations.

Training data

The data used to teach an AI model how language and facts look. For AEO, training data is one part of the pipeline: you want to be present in what models learn from, not just what they retrieve later.

Trustworthiness (in E-E-A-T)

Evidence that your content is accurate, fair, transparent, and safe to rely on. Includes clear sourcing, up-to-date information, disclosures about AI use, and alignment with relevant guidelines or regulations.

Vector database

A database built to store and search embeddings (vector representations of content). Used in many RAG systems to quickly find semantically similar passages or documents.

Voice search

Search conducted through spoken queries via phones, smart speakers, or in-car systems. Voice search often returns a single spoken answer, making AEO especially important.

Zero-click result

A search result where the user finds their answer directly on the SERP (for example, in an SGE panel, featured snippet, or knowledge box) without clicking through. AEO is designed to win in these scenarios by being the answer that's shown.

About the Authors

Zachary Handa is a search strategist and systems designer specializing in end-to-end visibility architecture for AI and search. His background spans a decade in digital marketing, including agency work, affiliate program management, and enterprise SEO product leadership at AT&T.

Zach specializes in building end-to-end visibility systems that combine entity management, structured content design, and AI-driven workflows. His work focuses on unifying human expertise with machine-readable structure, enabling brands to become trusted, retrievable sources within the evolving landscape of generative search.

Georg Sinn is the Founder of AllYouCanGET and a Senior Data Solution Architect who bridges the gap between complex data infrastructure and digital visibility.

With over two decades of experience in Business Intelligence and Data Governance at enterprises like IBM and Kyndryl, Georg leverages deep technical expertise in Software Development, Data Management Platforms like Databricks, and Data Integration to drive data excellence. He develops digital strategies, custom communication tools, and helps businesses scale through technology. His work focuses on the intersection of technical precision and societal impact, positioning him as a key expert in the structural data requirements essential for the emerging landscape of Answer Engine Optimization.

Made in the USA
Monee, IL
13 September 2026

62981871R00079